4/67

A FIRST COURSE IN
MATHEMATICAL ANALYSIS

A FIRST COURSE IN
MATHEMATICAL
ANALYSIS

BY

J. C. BURKILL

Sc.D., F.R.S.

*Fellow of Peterhouse and
Reader in Mathematical Analysis in the
University of Cambridge*

CAMBRIDGE
AT THE UNIVERSITY PRESS
1962

PUBLISHED BY
THE SYNDICS OF THE CAMBRIDGE UNIVERSITY PRESS

Bentley House, 200 Euston Road, London, N.W. 1
American Branch: 32 East 57th Street, New York 22, N.Y.
West African Office: P.O. Box 33, Ibadan, Nigeria

©

CAMBRIDGE UNIVERSITY PRESS
1962

Printed in Great Britain at the University Press, Cambridge
(Brooke Crutchley, University Printer)

CONTENTS

7. THE INTEGRAL CALCULUS

8. FUNCTIONS OF SEVERAL VARIABLES

PREFACE

This course of analysis is intended for students who have a working knowledge of the calculus and are ready for a more systematic treatment. Only a quite exceptional mathematician will then be mature enough for an axiomatic development of analysis in metric spaces, and he can be left to teach himself. The others normally follow a straightforward course based on the idea of a limit, and this book is an attempt to provide such a course. I have stopped short of Cauchy sequences, upper and lower limits, the Heine–Borel theorem and uniform convergence; in my experience many men understand those topics more readily if they are left to the next stage.

I am indebted to Professor G. E. H. Reuter and to Dr H. Burkill for their careful scrutiny of the manuscript.

J. C. B.

September 1961

1

NUMBERS

1.1. The branches of pure mathematics

This is a text-book of *mathematical analysis*. It is necessary first to say what is included under this heading. To this end we start with a short survey of the branches of pure mathematics. We are not concerned with mechanics or any other application of mathematics to natural science.

Mathematics as taught to the middle and upper forms of schools includes arithmetic, algebra, geometry, trigonometry and the calculus. No hard and fast boundaries are set up between these subjects and to solve a problem a student may employ ideas and methods from any of them.

A distinguishing feature of the calculus is that it rests on limiting processes. The gradient of a curve at a point P is the *limit* of the slope of a chord PQ as Q approaches P along the curve. In symbols, if the equation of the curve is $y = f(x)$, then the gradient is the derivative, dy/dx or $f'(x)$, defined by

$$f'(x) = \lim_{h \to 0} \frac{f(x+h) - f(x)}{h}.$$

The integral calculus also rests on the notion of limit. A basic problem of it is the calculation of an area bounded by a curved line. The only way in which such an area can be evolved from the areas defined in geometry is as the limit of the areas of polygons which approach the curve.

The idea of a limit is also encountered in the chapter of algebra on progressions, where it is seen that certain geometric progressions can be summed to infinity. In a well-defined sense which is easy to grasp, the unending series

$$\tfrac{1}{2} + \tfrac{1}{4} + \tfrac{1}{8} + \dots,$$

whose nth term is 2^{-n}, has sum 1. This means that we can make the sum of n terms as near to 1 as we like by taking n to be a sufficiently large number.

The notion of a limit rests on that of a *function*. The curve of which we were finding the gradient was specified by the function $f(x)$. The sum s_n of the first n terms of the geometric progression is expressed as a function of n by

$$s_n = 1 - 2^{-n}.$$

The idea of a function in its turn rests on that of *number*. The equation $y = f(x)$ of the curve expresses a connection between the number x and the number y. The sum s_n of the geometric progression depends on the number n (which does not vary continuously as the x can do but is restricted to be a positive integer).

1.2. The scope of mathematical analysis

We can now describe mathematical analysis as including those topics which depend on the notion of a limit. Thus it includes the differential and integral calculus, and you may ask whether a new title is necessary; does not *the calculus* adequately specify the subject matter? In a sense it does, and it is mainly by usage and tradition that *analysis* has come to denote a rather more formal (or more 'advanced') presentation, with greater attention to the foundations and more insistence on logical deduction. The use of the word analysis has the advantage of clearly including the summation of infinite series (which the schoolboy would reasonably regard as algebra rather than calculus).

Operations which are complete in a finite number of steps, such as the evaluation of a determinant, belong to algebra, not to analysis. The binomial theorem is a theorem of algebra if the index is a positive integer; otherwise it belongs to analysis.

Geometry is a subject, separate from analysis, developed from its own axioms. Its only impact on analysis is that we shall often find it suggestive and helpful to use geometrical language and illustrations.

In the light of what we have said the subject trigonometry is seen to fall into two parts. The solution of triangles, 'height and

distance problems' and the properties of trigonometric functions needed for them form a kind of practical geometry.
Results like

$$\sin x = x - \frac{x^3}{3!} + \frac{x^5}{5!} - \cdots,$$

which every one will recognise as being more exciting, belong
to analysis. After a first course in trigonometry, the viewpoint
must be changed. The sine and other trigonometric functions,
originally defined as ratios of lengths of lines, are seen to be
highly important functions of analysis, and the $\sin x$ should be
defined in terms of the variable x by the infinite series as it will
be in chapter 6 of this book.

Some knowledge of the trigonometric functions (and the exponential and logarithmic functions too) will be useful in earlier
chapters, but for the sole purpose of giving variety to the
examples. All references to the functions before chapter 6 could
be removed without affecting the sequence of theorems.

1.3. Numbers

We have seen that the logical order of development of mathematical analysis is

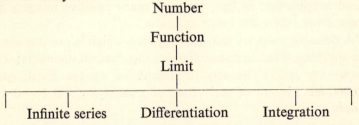

where the blank spaces in the last line can be filled by limiting
processes other than those already mentioned.

The first topic for investigation is number. When treated
exhaustively this is a difficult subject, with problems which have
roots both in mathematics and in philosophy. As this is a *first*
course in analysis we shall keep the discussion of number as
simple as we can, so long as it gives a firm foundation for the
structure of later definitions and theorems that will be set upon
it. The reader who wishes to go more deeply into the idea of

number may consult H. A. Thurston, *The Number-System* (Blackie, 1956); or E. Landau, *Foundations of Analysis* (Chelsea Publishing Co., 1951).

Sets. Before embarking on a discussion of number, we must say what is meant by a *set*. We often have to envisage all those persons or things having some assigned characteristic in common. Illustrations are: (i) all males of British nationality who are at a given time at least 18 years and less than 60 years old, (ii) all mountain-tops on the Earth over 10,000 feet high, (iii) all positive integers, (iv) all equilateral triangles in a given plane. Such collections determined by some defining property we shall call *sets*. The words *class* and *aggregate* are also used with the same meaning. We emphasise that a set is known without ambiguity whenever the rules defining it enable us to say of any proposed candidate whether it is or is not a member of the set. For instance, some readers are included in the set (i) and others are not, but the rules are clear and no one is left in doubt.

The examples (i)–(iv) illustrate the distinction between *finite* and *infinite* sets. The sets (i) and (ii) are finite; with sufficient knowledge and patience a complete list could be provided of the members of each of them. The sets (iii) and (iv) on the other hand are infinite; in (iii), however many positive integers we write down, there are more to follow.

A defining property may be proposed which is not possessed by anything. The corresponding set then has no members; it is *empty* (or *null*). The sets of mountains on the Earth over 30,000 feet high or of real values of x satisfying $x^2 + 1 = 0$ are empty.

Integers. We take for granted the system of positive integers

$$1, 2, 3, \ldots,$$

stressing only those facts which are the most important for further extensions of the number-system.

Positive integers a, b can be added or multiplied and there are positive integers c, d such that

$$a + b = c \quad \text{and} \quad ab = d.$$

The integer 1 has the property that, for every positive integer a,

$$1.a = a.1 = a.$$

The integers have an *order* expressed by $<$ or $>$.
The letter n will always denote a positive integer.

The principle of induction. *If a statement* $P(n)$ *is*
  (i) *true for* $n = 1$,
  (ii) *true for* $n+1$ *whenever it is true for* n,
then it is true for every positive integer n.

The principle of induction is often useful as a method of proof (see exercises 1(a)).

We have mentioned addition and multiplication of positive integers; we turn to subtraction, and afterwards to division.

In the system of positive integers the equation

$$a+x = b$$

can be solved for x only if $a < b$. If it is to have a solution when $a = b$ or $a > b$ we must introduce zero and the negative integers. We shall then have widened our number-system to contain all integers, which can be arranged in order

$$..., \ -3, \ -2, \ -1, \ 0, \ 1, \ 2, \ 3, \ ....$$

Rational numbers. If a and b are integers, the equation

$$bx = a$$

is not in general satisfied by an integral value of x. If this equation is always to have a solution (b not being 0) we must widen the system to include rationals a/b. In the system of rationals the operations of arithmetic are straightforward and familiar to the reader.

A relation of *order* naturally suggests itself for the rationals. Supposing that b and d are positive integers we define

$$\frac{a}{b} > \frac{c}{d}$$

to mean $ad > bc$.

Between any two rationals there is another (and, hence, infinitely many others).

To prove this, we remark that, if b and d are positive integers, the rational

$$\frac{a+mc}{b+md}$$

lies between a/b and c/d for any positive integer m.

We may describe this property of the rationals by saying that they are *dense* in any interval.

Exercises 1 (*a*)

Notes on these exercises are given on p. 170.

The method of induction may be used for **1–6**.

1. $\sum\limits_{r=1}^{n} r^2 = \frac{1}{6}n(n+1)(2n+1)$, where $\sum\limits_{r=1}^{n} r^2$ means $1^2+2^2+\ldots+n^2$.

2. $1^2+3^2+5^2\ldots+(2n-1)^2 = \;?$

3. $1.1+3.2+5.2^2+\ldots+(2n+1)2^n = A+(B+Cn)2^n$, where A, B and C are constants (not depending on n) to be found.

4. $\sum\limits_{r=1}^{n} \dfrac{x^{2^{r-1}}}{1-x^{2^r}} = \dfrac{1}{1-x} - \dfrac{1}{1-x^{2^n}}.$

5. $2^n > n^3$ if $n > 9$.

6. $5^{2n}-6n+8$ is divisible by 9.

7. If b, d, q are positive integers and

$$\frac{a}{b} > \frac{p}{q} > \frac{c}{d},$$

prove that positive integers m, n can be found such that

$$\frac{p}{q} = \frac{ma+nc}{mb+nd}.$$

Construct a numerical example and solve it.

8. The density property of the rationals amounts to saying that there is no rational which is *next* to another. Observe the following plan of arranging the positive rationals (not in order of magnitude) which does assign a definite place to each

$$\tfrac{1}{1}; \; \tfrac{2}{1}, \tfrac{1}{2}; \; \tfrac{3}{1}, \tfrac{2}{2}, \tfrac{1}{3}; \; \tfrac{4}{1}, \tfrac{3}{2}, \tfrac{2}{3}, \tfrac{1}{4}; \; \tfrac{5}{1}, \ldots.$$

Prove that p/q occupies the $\{\frac{1}{2}(p+q-1)(p+q-2)+q\}$th place. (Each rational occurs infinitely often; e.g. 1 appears as $\tfrac{1}{1}, \tfrac{2}{2}, \tfrac{3}{3}, \ldots.$)

1.4. Irrational numbers

It was realised by the Greeks more than 2000 years ago that there is an incompleteness about the system of rational numbers. The diagonal of a square with sides of unit length has a length which is *irrational*. In algebraic language, the equation for x

$$x^2 = a$$

has a rational solution only for exceptional values of the rational number a (for example, 4 or 4/9) and is not so soluble if, say, a is 2 or 3 or 5/9.

The first theorem of the book will be a formal proof that the square root of 2 is irrational. According to our criterion of §1.2, it is really a theorem of algebra rather than of analysis. But it earns its place, first, by its historic interest—it was proved by Pythagoras or one of his school—and secondly, by the neatness and economy of its argument.

Theorem 1.4. *No rational number has square* 2.

Proof. Suppose, on the contrary, that the rational a/b has square 2, where a and b are integers having no common factor. Then

$$a^2 = 2b^2.$$

Since 2 divides a^2, the integer a must be even.

Write $a = 2c$, where c is an integer. Then

$$2c^2 = b^2.$$

Then 2 divides b^2 and so b must be even.

Thus a and b both have the factor 2, which contradicts the hypothesis. **|** (We use this thick vertical stroke sometimes to denote that the proof of a theorem is complete.)

We have so far presented only the simplest specimens of numbers which are not rational. We add others which are less simple.

(i) x is the positive number which satisfies the equation

$$x^3 = x + 7.$$

(It is possible to prove that there is just one such x.) By methods of the theory of equations, x can be expressed in terms of cube roots of rational numbers.

(ii) x is the positive number which satisfies the equation

$$x^5 = x + 7.$$

We might expect that x could be represented by some combination of roots of rational numbers, perhaps fifth roots. But this is not so. A difficult theorem of algebra shows that roots of equations of degree higher than four cannot generally be so expressed.

(iii) The number π, the ratio of the circumference of a circle to its diameter.

A method of proving that π is irrational is outlined in exercise $7(f)$, 7. It can be proved (by a more difficult argument) that π does not satisfy any algebraic equation with integer coefficients. So it is a number which is, in a sense, even less easy to grasp than those in (i) and (ii).

Exercises 1 (*b*)

Notes on these exercises are given on p. 170.

1. Adapt the argument of theorem 1.4 to show that no rational number has its cube equal to 16.

2. Extend the result of **1** to show that a rational number p/q in its lowest terms can be the cube of a rational number only if p and q are cubes of integers.

3. Prove the more general theorem (Gauss, 1777–1855) that, if $p_1, p_2, \dots, p_n$ are integers, the only possible rational roots of the equation

$$x^n + p_1 x^{n-1} + p_2 x^{n-2} + \dots + p_n = 0$$

are integers which divide p_n.

4. Solve the equations:

$$x^4 - x^3 - 16x^2 + 4x + 48 = 0,$$
$$4x^3 - 8x^2 - 3x + 9 = 0.$$

1.5. Cuts of the rationals

In §1.4 we showed the need of completing the number-system by 'filling the gaps' which occur among the rationals. It is possible to give different constructions for filling the gaps; we follow the procedure of Dedekind (1872).

Before stating it in general terms, we think it helpful to show how a particular irrational number, say $\sqrt{2}$, is fitted in among the rationals.

In trying to isolate a number whose square is 2, we first observe from theorem 1.4 that the positive rational numbers fall into two classes, those whose squares are less than 2 and those whose squares are greater than 2. Call these classes the left-hand class L and the right-hand class R, corresponding to their relative positions when represented graphically on a horizontal line. Examples of numbers l in L are 7/5 and 1·41, and of numbers r in R are 17/12 and 1·42. The reader will convince himself that any r is greater than any l and—with a little more thought—that there is no member l of L which is greater than all the other members and there is likewise no r which is the least member of R.

The statements in the last paragraph become more concrete if we use the arithmetical rule for square root to find, to as many decimal places as we please, a set of numbers l

$$1, \ 1\cdot4, \ 1\cdot41, \ 1\cdot414, \ 1\cdot4142, \ ...,$$

each of which is greater than the preceding (or equal to it if the last digit is 0) and each having its square less than 2. Moreover, the numbers got by adding 1 to the last digit of these numbers l form a set of numbers r

$$2, \ 1\cdot5, \ 1\cdot42, \ 1\cdot415, \ 1\cdot4143, \ ...,$$

each having its square greater than 2 and each less than (or equal to) the preceding.

If now we are given a particular rational number a whose square is less than 2, we shall by going far enough along the set of numbers 1, 1·4, 1·41, 1·414, ... come to one which is greater than a. (Alternatively, this can be proved by the method of exercise 1 (c), 1.)

If, then, we are building up a number-system starting with integers and then including the rational numbers, we see that an irrational number (such as $\sqrt{2}$) corresponds to and can be defined by a cutting of the rationals into two classes L, R of which L has no greatest member and R no least member. This is Dedekind's definition of irrationals by the *cut*.

Exercises 1 (c)

Notes on these exercises are given on p. 170.

1. Prove that, if m/n is an approximation to $\sqrt{2}$ from below, then $(m+2n)/(m+n)$ is a closer approximation from above. Hence write down approximations to $\sqrt{2}$, obtaining two which differ by less than $1/10,000$.

2. Find similarly approximations to $\sqrt{3}$.

3. Prove that, if a, b, c, d are rational and

$$a+\sqrt{b} = c+\sqrt{d},$$

then either (i) $a = c$, $b = d$, or (ii) b and d are both squares of rational numbers.

4. Prove that, if a, b, c are rational and

$$a+b\sqrt{2}+c\sqrt{3} = 0,$$

then

$$a = b = c = 0.$$

5. If a, b, c are rational and

$$a+b\sqrt[3]{2}+c\sqrt[3]{4} = 0,$$

what conclusion can you draw?

6. If a, b, c, d are rational and x is irrational, in what circumstances is

$$\frac{ax+b}{cx+d}$$

rational?

1.6. The field of real numbers

In §§1.3–1.5, starting from integers, we have sketched the building-up of the system of real numbers. To fill in all the detail could be a term's work. We should have to prove that the numbers obey the familiar rules of algebra, of which

$$a(b+c) = ab+ac$$

is one instance. Any reader who would like to amplify this cursory treatment of the subject should consult one of the books mentioned in §1.3.

Our plan at this stage is to make a list of the basic properties which the real numbers satisfy. As we offer no proof of these properties, we treat them as axioms. They fall naturally into three sets covering respectively algebraic manipulation, order and completeness.

The reader who is studying modern algebra at a similar level to this course of analysis will find there that the first set of axioms are those which define a *field*.

A system, in algebra, means a set of things or elements, together with operations on them. A field, denoted by F, is by definition a system whose elements $a, b, c, \ldots$ are subject to two operations $+$ and $\times$, satisfying the following algebraic axioms A1–11:

A1. Every two elements a, b in F have a *sum* $a+b$ in F.

A2. $a+b = b+a$.

A3. $(a+b)+c = a+(b+c)$.

A4. There is an element 0 in F such that $0+a = a$ for every a.

A5. For every a in F there is x in F such that $a+x = 0$. We write $-a$ for this (unique) x.

The axioms A6–10 which follow are the analogues for the operation $\times$ of A1–5 for $+$.

A6. Every two elements a, b in F have a *product* $a \times b$ in F. Following ordinary usage we can generally shorten $a \times b$ into ab.

A7. $ab = ba$.

A8. $(ab)c = a(bc)$.

A9. There is an element 1 in F such that $1a = a$ for every a.

A10. For every a in F except 0 there is y in F such that $ay = 1$. We write $1/a$ for this (unique) y.

The final axiom A11 links the two operations $+, \times$.

A11. $(a+b)c = ac+bc$.

From the axioms A1–11 the familiar rules of manipulation of real numbers can be deduced. As illustrations, proofs of two of them are appended.

(i) *The cancellation law.* If $ab = ac$ and $a \neq 0$, then $b = c$.

Proof.
$$b = \left(\frac{1}{a}\,a\right)b \quad \text{(A10)}$$

$$= \frac{1}{a}(ab) \quad \text{(A8)}$$

$$= \frac{1}{a}(ac) \quad \text{(given)}$$

$$= \left(\frac{1}{a}\,a\right)c \quad \text{(A8)}$$

$$= c.$$

(ii) '*Two minuses make a plus*'. $(-a)(-b) = ab$.

Proof. $ab + a(-b) = a(b + (-b))$ (A11)

$$= a0 = 0.$$

Similarly $(-a)(-b) + a(-b) = (-a + a)(-b) = 0.$

From A5, $ab = (-a)(-b).$

In a general field there is no relation of order by which we can say that, of any two elements, one precedes the other. Since the field of real numbers does possess an ordering relation (that of $>$), we add now the relevant axioms O for an *ordered field*:

O1.   For every a, b in F one and only one of

$$a > b, \quad a = b, \quad b > a$$

is true.

O2.   If $a > b$ and $b > c$, then $a > c$.

O3.   If $a > b$, then $a + c > b + c$.

O4.   If $a > b$ and $c > 0$, then $ac > bc$.

From our knowledge of the rational numbers we see that the axioms A and O are appropriate to them and that they form an ordered field. Observe that the integers do not form a field, because they do not satisfy A10.

To lay down a set of axioms for the real numbers, as distinct from the rationals, we must add one expressing *completeness*, in the sense that the gaps among the rationals are filled.

The axiom of completeness, which we shall state in the form due to Dedekind, is necessarily more intangible and abstract than the axioms A and O. You should read it now and turn back to study it again when it is used in proofs of theorems (e.g. 1.8).

Dedekind's axiom. *Suppose that the system of all real numbers is divided into two classes L, R, every member l of L being less than every member r of R (and neither class being empty). Then there is a dividing number ξ with the properties that every number less than ξ belongs to L and every number greater than ξ belongs to R. The number ξ itself may belong either to L or to R. If it is in L, it is the greatest member of L; if it is in R, it is the least member of R.*

Such a division of the real numbers into two classes by means of some rule is called a *Dedekind cut*.

1.7. Bounded sets of numbers

Consider the following sets of real numbers:
(1) All prime numbers.
(2) All positive integers less than 1000.
(3) All integers greater than 1000 which are perfect squares.
(4) All rational numbers x such that $1 \leqslant x \leqslant 3$.
(5) All real numbers x such that $1 \leqslant x \leqslant 3$.
(6) All real numbers x such that $1 < x < 3$.

Observe that these sets are infinite with the exception of (2) which has a finite number, 999, of members. Examples (1) and (3) might give an impression that an infinite set has to contain members which are large numbers; examples (4) to (6) would correct this false impression.

The sets (5) and (6) are of an important and simple type, and each is called an *interval*. The former, in which the *end-points* 1 and 3 are members of the set, is a *closed* interval; and the latter, in which the end-points are excluded from the set, is an *open* interval. An interval $a \leqslant x \leqslant b$ or $a < x < b$ will often be written (a, b).

Some writers use distinctive notations which show whether an interval is open or closed, e.g. (a, b) for an open interval and $[a, b]$ for a closed interval. We shall not adopt any such convention in this book, but the reader may do so if he wishes.

An interval (a, b) will be called a finite interval. The set of x for which $x \geqslant a$ form an infinite interval.

The greatest and least numbers of a set. If S is a set consisting of finitely many different real numbers, plainly there is one member of the set which is greater than all the others and one member which is less than all the others. Convenient abbreviations for these greatest and least numbers are *max* and *min*.

Illustration. If S consists of all three-figure even integers, max S is 998 and min S is 100.

If now S is a set with infinitely many members, there may or may not be a member of S which is greater than all the others (or one which is less than all the others). We illustrate by examples.

(1) If S is the closed interval $(-1, 1)$, i.e. the set of x for which $-1 \leqslant x \leqslant 1$, then the number 1 is greater than all the other members of S.

(2) If S is the open interval $(-1, 1)$, i.e. the set of x for which $-1 < x < 1$, there is no member of S which is the greatest. If k is any member of S, then $k < \frac{1}{2}(1+k) < 1$ and $\frac{1}{2}(1+k)$ is a member of S greater than k.

(3) If S is the set of integers which are perfect squares, i.e. 1, 4, 9, ... there is a least member but no greatest member.

Definitions. Let S be a set of real numbers. If there is a number K such that, for every member x of S,

$$x \leqslant K,$$

we say that S is *bounded above.* K is called *an upper bound* of S. Similarly, if there is a k such that $x \geqslant k$ for every x in S, then S is *bounded below*, and k is *a lower bound* of S.

If S is bounded both above and below we say simply that it is *bounded.* A set which is not bounded is called *unbounded.*

Illustrations. (1) Any finite set S is bounded; and max S, min S can be taken as upper and lower bounds.

(2) The set of numbers,

$$\frac{1}{2}, \frac{2}{3}, \frac{3}{4}, \ldots, \frac{n}{n+1}, \ldots,$$

where n takes all positive integral values, is bounded. The number $\frac{1}{2}$ (or any smaller number) serves as a lower bound, and 1 (or any greater number) as an upper bound. Note carefully that the set has no greatest member.

(3) The set $-1, -\sqrt{2}, -\sqrt{3}, \ldots, -\sqrt{n}, \ldots$ is bounded above but is not bounded below.

If K is an upper bound of a set S, then any number greater than K is also an upper bound. If all that we desire to assert is the boundedness of a set, one upper bound is as good as another. If we want to make the sharpest possible statement, confining the set as closely as we can, we shall aim at choosing the *least upper bound*, i.e. a number K which is an upper bound but such that $K-\epsilon$ (where ϵ is any positive number however small) is exceeded by some member of the set S. Similarly we should seek the *greatest lower bound.*

We shall prove in §1.8 that, if a set S is bounded above, it is always possible to make this most economical choice of an upper bound.

Illustration. In (2) above, 1 is the least upper bound. Any number less than 1 is exceeded by the member $n/(n+1)$ of the set if n is large enough (e.g. the number 0·99, less than 1, is exceeded by $n/(n+1)$ when $n > 99$).

1.8. The least upper bound (supremum)

The next theorem is one of the foundation-stones of analysis and, in any orderly development of the subject, it must be found near the beginning. The reader should master its meaning and should test its truth by constructing for himself examples (such as those at the end of this section). If he finds the proof, based on Dedekind's axiom, natural and comprehensible, so much the better. If he finds it more difficult to follow than the arguments that he has so far encountered in mathematics, he need not be disheartened, but should read the succeeding chapters and return later to a study of the foundations.

Theorem 1.8. *If S is a (not-empty) set of numbers which is bounded above, then of all the upper bounds there is a least one.*

Proof. Divide the real numbers x into two classes L, R by these rules.

Put x in L if there is a member s of S such that $s > x$.

Put x in R if, whatever member s of S is taken, $s \leqslant x$.

Then every x goes either into L or into R. Moreover, neither L nor R is empty. For, if s is some member of S, then (say) $x = s-1$ is in L. And, since S is bounded above, any upper bound K, for which $s \leqslant K$ for all s, is in R.

Any l of L is less than any r of R. For there is some s which is greater than l, and this s is less than or equal to r.

By Dedekind's axiom there is a dividing number ξ such that, for every positive ϵ, $\xi - \epsilon$ is in L and $\xi + \epsilon$ in R. In the Dedekind axiom, ξ itself may belong either to L or to R. We shall prove that, in the present application, ξ belongs to R.

Suppose, if possible, that ξ belongs to L. Then there is a member s of S with $s > \xi$.

The number $\eta = \frac{1}{2}(s+\xi)$ satisfies $s > \eta > \xi$: η is in R since it is greater than the dividing number ξ. So $s \leqslant \eta$ by the rule for R. This contradicts our earlier inequality $s > \eta$. So ξ belongs to R.

We have proved that ξ satisfies

(1) $s \leqslant \xi$ for every s in S.

(2) $\xi - \epsilon$ being any number less than ξ, there is an s for which $s > \xi - \epsilon$.

The property (1) shows that ξ is an upper bound of S, and (2) that it is the least upper bound. The theorem is proved.

Illustrations. (1) Let S be the rational numbers x for which $0 \leqslant x \leqslant \frac{1}{2}$. Then $\frac{1}{2}$ is the least upper bound. It is also the greatest member of S.

(2) Let S be the rational numbers x for which $x^2 < 2$. The number $\sqrt{2}$ is the least upper bound.

The least of the upper bounds of a set is so vitally related to it as to merit a name of its own. Some writers call it *the* upper bound (distinguished from the indefinite 'an upper bound'); others use the initials l.u.b. A term which is expressive and concise is *supremum*, abbreviated *sup*. We can sum up as follows.

Definition. *If, given a set of numbers S, there is a number K such that*

(1) *$s \leqslant K$ for every s in S,*

(2) *for every positive ϵ, there is an s in S for which*

$$s > K - \epsilon,$$

then we write $K = \sup S$.

Theorem 1.8 proved the existence of K when the set S is bounded above.

By reversing inequality signs, we set up an analogous theory of lower bounds and the greatest of the lower bounds (the *infimum*, abbreviated *inf*).

Definition. If, given a set of numbers S, there is a number k such that

(1) $s \geqslant k$ for every s in S,

(2) for every positive ϵ, there is an s in S for which

$$s < k + \epsilon,$$

then we write $k = \inf S$.

A set S which is bounded below can be proved to have an infimum.

Exercises 1 (*d*)

Notes on these exercises are given on pp. 170–1.

Note. The relations of inequality in exercises 5–8 are repeatedly useful in analysis and it pays to be familiar with them. They may already be known to the reader as results in algebra. Proofs are given on pp. 170–1.

1. If a and b are numbers, and

$$a < b + \frac{1}{n}$$

for every positive integer n, prove that $a \leqslant b$.

The same is true if, in place of $1/n$, we write ϵ, where ϵ can take every value greater than 0.

2. Let A and B be two bounded sets of real numbers. Let $A \cup B$ denote the set of all numbers which are in A or in B (or in both). Prove that

$$\sup (A \cup B) = \max (\sup A, \sup B).$$

Is there any corresponding result for $A \cap B$, the set of numbers which are in both A and B?

3. Let A and B be bounded sets of real numbers. Let C be the set of all numbers c, where $c = a + b$ and a is any member of A and b any member of B. Prove that

$$\sup C = \sup A + \sup B.$$

Is there a corresponding result for the set D of numbers d where $d = ab$?

4. What are the sup and inf of the set of numbers $2^{-m} + 3^{-n}$, where m and n take all positive integral values?

5. (*The inequality of the arithmetic and geometric means.*) If $a_1, a_2, ..., a_n$ are positive and

$$A = \frac{a_1 + a_2 + ... + a_n}{n}, \quad G = (a_1 a_2 ... a_n)^{1/n},$$

then

$$A \geqslant G,$$

with equality if and only if the a_r are all equal.

6. (*Cauchy's inequality.*) For any two sets of real numbers $a_1, a_2, ..., a_n$ and $b_1, b_2, ..., b_n$,

$$(\Sigma a_r b_r)^2 \leqslant (\Sigma a_r^2)(\Sigma b_r^2)$$

with equality if and only if there are constants k, l such that $k a_r = l b_r$ for all r (i.e. if the a_r and b_r are proportional).

7. If $a > 1$ and r, s are rationals with $r > s > 0$, then

$$\frac{a^r - 1}{r} > \frac{a^s - 1}{s}.$$

8. If $a < 1$ and r, s are as in **7**,

$$\frac{1 - a^r}{r} < \frac{1 - a^s}{s}.$$

1.9. Complex numbers

The system of real numbers is comprehensive enough to carry very many theorems of mathematical analysis. It may be asked what we should miss by not admitting at any stage numbers other than real numbers. In answering this question it is customary to point out that some quadratic equations, of which the simplest is

$$x^2 + 1 = 0,$$

have no roots in the system of real numbers. This is a deprivation rather than a disaster. It is indeed satisfying that the introduction of complex numbers enables us to prove the theorem that every algebraic equation has a root. But an even more cogent case for admitting complex numbers rests on their bringing to light close connections between some of the most common functions of analysis, the exponential function on the one hand and the trigonometric—sine and cosine—on the other. If the variables are real these functions are for ever unrelated. The case for including complex numbers in ordinary analysis is finally won by the beauty and generality of some of the later theorems which are based on them (beyond the scope of this course).

We now sketch a method of introducing complex numbers into analysis. We can extend the field of real numbers by *adjoining* one or more new elements which are combined with the original members of the field by the operations + and × in accordance with the axioms A. The new element which must be adjoined to yield complex numbers is i, where i by definition satisfies $i^2 + 1 = 0$. The numbers $a + bi$, where a and b are real, which are elements of the extended field are added and multiplied in accordance with the algebraic axioms A1–11. The number $a + 0i$ behaves in every respect like the real number a.

No axioms of the type O (order) hold in the extended field. It is not possible to arrange complex numbers in an order of magnitude in the way that real numbers can be so arranged.

Exercises 1 (*e*)

Notes on these exercises are given on p. 171.

1. Prove from the axioms that, if *a*, *b*, *c*, *d* are real and $a + bi = c + di$, then $a = c$ and $b = d$.

2. The inverse (A 10) of $a + bi$ is $(a - bi)/(a^2 + b^2)$ provided that *a* and *b* are not both 0.

3. Prove that, if the product of two complex numbers is zero, at least one of them is zero.

4. From the addition formulae for the cosine and sine deduce that

$$(\cos\theta + i\sin\theta)(\cos\phi + i\sin\phi) = \cos(\theta + \phi) + i\sin(\theta + \phi).$$

5. By induction or otherwise prove de Moivre's theorem

$$(\cos\theta + i\sin\theta)^n = \cos n\theta + i\sin n\theta.$$

6. Extend de Moivre's theorem taking the index to be (i) a negative integer, (ii) a rational p/q.

1.10. Modulus and phase

The usual notation for a variable complex number is $z = x + yi$, where *x* and *y* are real. We remarked in §1.2 that geometrical illustrations often help in analysis. The geometrical representation of complex numbers is particularly suggestive. Taking a plane with a pair of rectangular axes Ox, Oy, we represent the complex number $z = x + yi$ by the point whose coordinates are (x, y). The number *x* is called the real part and *y* the imaginary part of *z*, written

$$x = \text{re } z, \quad y = \text{im } z.$$

The main advantage of this representation is that the sum of two complex numbers z_1 and z_2 corresponds to the point *P* where the vector *OP* is the sum of the vectors from *O* to the points representing z_1 and z_2.

If $z = x + yi$, the positive number $r = \sqrt{(x^2 + y^2)}$ is called the *modulus* of *z*, written $|z|$. The angle θ such that $\cos\theta = x/r$ and $\sin\theta = y/r$ is the *phase* of *z*, ph *z*; some writers call it amplitude or argument. The angle θ is indeterminate in the sense that any multiple of 2π can be added to or subtracted from it. It is often convenient to have a *principal value* of the phase, and this is defined to be the value of θ such that $-\pi < \theta \leqslant \pi$. Thus

$$z = x + yi = r(\cos\theta + i\sin\theta).$$

The numbers $x+yi$ and $x-yi$ are called *conjugates* and denoted by z and $\bar{z}$. Observe that the sum and the product of two conjugate complex numbers are real, and also that

$$z\bar{z} = |z|^2.$$

Theorem 1.10. (*Modulus of product and sum*).
  If $z = x+yi$ and $w = u+vi$, then

(1) $|zw| = |z| \cdot |w|$.

(2) $|z+w| \leqslant |z|+|w|$,

and the sign $=$ holds if and only if the phases of z and w are the same (or differ by a multiple of 2π).

Observe that the geometrical counterpart of the sum-theorem is that one side of a triangle is less than the sum of the other two. We must of course give an analytical proof.

Proof. (1) To prove the statement about the product zw, we have
$$|zw|^2 = (zw)(\overline{zw}) = (z\bar{z})(w\bar{w}) = |z|^2|w|^2,$$

and taking the square root gives the result since all the moduli are positive.

(2) $\begin{aligned}|z+w|^2 &= (z+w)(\bar{z}+\bar{w}) \\ &= z\bar{z}+(z\bar{w}+w\bar{z})+w\bar{w} \\ &= |z|^2+2\,\mathrm{re}\,(z\bar{w})+|w|^2.\end{aligned}$

Now $\qquad -|zw| \leqslant \mathrm{re}\,(z\bar{w}) \leqslant |zw|$

and $\mathrm{re}\,(z\bar{w}) = |zw|$ if and only if $z\bar{w}$ is real and positive, i.e. if z and w have the same phase . So $|z+w| \leqslant |z|+|w|$, and the sign $=$ holds if and only if z and w have the same phase. |

Further extensions of the number-system? The reader may well ask whether it will be profitable to extend the notion of number beyond that of complex number. As the space in which we move has three dimensions it is tempting to suppose that the mathematical description of natural phenomena could make good use of numbers of the form $x+yi+zj$, where x, y, z are real cartesian coordinates and j is some element which, like i, can be adjoined to the field of real numbers. The answer turns out to be that, extensions, though possible, are not useful. The price to be paid in extra complexity and loss of desirable properties is too high. We already paid the price of sacrificing the order relation in the step from real to complex numbers. A further step would in fact lead us to the system of quaternions of the form $x+yi+zj+wk$; these numbers have some interesting properties but they have the heavy disadvantage of not obeying the commutative law of multiplication $ab = ba$.

Exercises 1 (f)

Notes on these exercises are given on p. 171.

1. Taking a simple value of z (such as $2+i$), mark in a diagram the points representing $z+1$, $2z$, iz, $1/z$, $\bar{z}$, $z+\bar{z}$, $z\bar{z}$.

2. What are the loci

$$|z-1| = k|z+1|+l$$

for the pairs of values $(k, l) = (0, 2)$, $(1, 0)$, $(2, 0)$, $(1, 1)$, $(1, 3)$? How is each locus related to the points $z = 1$, $z = -1$?

3. Solve the equations:

$$\text{(i)} \quad z^4 = 28+96i,$$

$$\text{(ii)} \quad z^6 - 3z^3 + 2 = 0,$$

$$\text{(iii)} \quad 2z^3 + \bar{z}^3 = 3.$$

4. Prove that the roots of

$$z^3 + 3pz^2 + 3qz + r = 0$$

form an equilateral triangle if and only if $p^2 = q$.

5. Prove that, if $|a_r| \leqslant 2$ for $1 \leqslant r \leqslant n$, then every root of

$$1 + a_1 z + \ldots + a_n z^n = 0$$

has modulus greater than $\frac{1}{3}$.

6. If a and c are real, what is the locus

$$az\bar{z} + b\bar{z} + bz + c = 0?$$

7. On the sides of a triangle $Z_1 Z_2 Z_3$ are constructed isosceles triangles $Z_2 Z_3 W_1$, $Z_3 Z_1 W_2$, $Z_1 Z_2 W_3$, lying outside the triangle $Z_1 Z_2 Z_3$. The angles at W_1, W_2, W_3 are all $2\pi/3$. Prove that the triangle $W_1 W_2 W_3$ is equilateral.

8. Let $a \neq 0$ and $a\bar{a} \neq c\bar{c}$. Prove that a root of

$$az^2 + bz + c = 0$$

has modulus 1 if and only if

$$|\bar{a}b - \bar{b}c| = |a\bar{a} - c\bar{c}|.$$

9. If v is a complex root of $x^7 - 1 = 0$, express the other six roots in terms of v. Find a quadratic with real coefficients having a root $v + v^2 + v^4$. Prove that

$$\cos \frac{\pi}{7} - \cos \frac{2\pi}{7} + \cos \frac{3\pi}{7} = \frac{1}{2},$$

$$-\sin \frac{\pi}{7} + \sin \frac{2\pi}{7} + \sin \frac{3\pi}{7} = \tfrac{1}{2}\sqrt{7}.$$

In exercises 10–13, $P(z)$ denotes a polynomial in z

$$a_0 z^n + a_1 z^{n-1} + \ldots + a_n,$$

where the coefficients a , ..., a_n are complex (unless they are given to be real).

10. Prove that $P(z) = A + Bi$, where A and B are polynomials in x and y with real coefficients.

11. Prove that a rational function $R(z)$, defined as the quotient of two polynomials, can be reduced to the form $X + Yi$, where X and Y are rational functions of x and y with real coefficients.

12. If the coefficients of powers of z in $R(z)$ are real and

$$R(x + yi) = X + Yi,$$

prove that $R(x - yi) = X - Yi.$

13. If the coefficients a , ..., a_n in $P(z)$ are real, prove that the roots of the equation $P(z) = 0$ are real or consist of conjugate pairs.

2

SEQUENCES

2.1. Sequences

In chapters 2–4 we shall operate with real numbers only. Complex numbers will be needed in chapter 5. A *sequence* is a set of numbers occurring in order, that is to say, there is a first number, a second number and so on. If the sequence is unending, or, in other words, if, whatever positive integer n be assigned, there is a corresponding nth number, we have an *infinite sequence*. In simple cases a sequence is defined by an explicit formula giving the nth number in terms of n. The nth number of a sequence is conveniently denoted by s_n (or t_n or u_n, etc.).

Illustrations.

(1) $s_n = 1/n$. The sequence is $1, \frac{1}{2}, \frac{1}{3}, \ldots$.

(2) $s_n = (-1)^n/\sqrt{n}$. (3) $s_n = n^2$.

Sequences provide the easiest introduction to the idea of limit, which, as we said, is fundamental in mathematical analysis. We discuss particular examples as a preparation for the ensuing formal presentation.

2.2. Null sequences

In illustration (1), the nth member (or term) of the sequence, namely $1/n$, becomes smaller as n becomes larger, and, by taking n large enough, we can make s_n as close as we like to zero. To take a numerical illustration, s_n is less than $0\cdot0001$ for every integer n greater than 10^4. Such a sequence is called a *null sequence*. The illustration (2) gives another null sequence; the numbers do not decrease with every step from n to $n+1$ as those of (1) do, but the requirement of arbitrarily close approach to zero is fulfilled. You are now ready for a precise statement.

Definition. s_n *is a* null sequence *if, to every positive number* ϵ, *there corresponds an integer N such that*

$$|s_n| < \epsilon \quad \text{for all values of n greater than N.}$$

You should study this definition with care and frame examples by which to test it. Observe that the criterion for a null sequence may be displayed by writing values of ϵ and N in two columns. In the above illustrative example (2), entries in the columns might be

ϵ	N
0·001	10^6
0·00001	10^{10}

and, in fact, we can make the rule that N can be chosen to be any integer not less than $1/\epsilon^2$. It is not necessary to assign the smallest possible value of N. The symbol ϵ (and sometime δ or η) is an established notation for a small positive number. It is to be assumed without explicit mention that $\epsilon > 0$.

You should decide of each of the following sequences whether or not it is a null sequence,

(4) $\sin n\pi$,   (6) $n^{-1} \sin \tfrac{1}{4}n\pi$,

(5) $\sin \tfrac{1}{2}n\pi$,   (7) $n/(n^2+1)$.

The following points should be noted.

(*a*) The alteration of the values of s_n for any finite number of values of n will not affect the question whether it is a null sequence. Suppose, for instance, that $s_n = 1/(n-10)$. Then s_n is not defined for $n = 10$, and the natural course would be to start the sequence at $n = 11$. It is a null sequence.

Alteration of values for infinitely many n would, however, make an essentially different sequence. For example, suppose that $s_n = 1/n$ for all values of n except powers of 2 and $s_n = 1$ for $n = 2, 4, 8, 16, \ldots$; then s_n is not a null sequence.

(*b*) A null sequence may or may not actually take the value zero. The two possibilities are illustrated by two of the preceding examples.

(1) $s_n = 1/n$. No number s_n is equal to zero.
(6) $s_n = n^{-1} \sin \tfrac{1}{4}n\pi$. $s_n = 0$ when n is a multiple of 4.

2.3. Sequence tending to a limit

A null sequence is one whose terms approach zero. It is easy to adapt the definition to a sequence whose terms approach any number s.

Illustration. The numbers of the sequence $\frac{1}{2}, \frac{2}{3}, \frac{3}{4}, \ldots, n/(n+1), \ldots$ approach the value 1.

Definition. A sequence s_n is said to tend to the limit s *if, given any positive ϵ, there is N such that*

$$|s_n - s| < \epsilon \quad \text{for all } n > N.$$

We then write $\qquad\qquad \lim s_n = s.$

Notes.

(1) Clearly, $\lim s_n = s$ if and only if $s_n - s$ is a null sequence.

(2) The inequality $|s_n - s| < \epsilon$ is equivalent to the two inequalities

$$s - \epsilon < s_n < s + \epsilon.$$

This expanded form is often clearer.

(3) There is a short and expressive notation, the arrow,

$$s_n \to s$$

meaning $\qquad\qquad \lim s_n = s.$

(4) There is a further symbolism which saves much writing and which you may adopt when (but not before) you have mastered its meaning. The above definition may be written

$s_n \to s$ if $\quad \epsilon > 0; \quad \exists \, N. \quad |s_n - s| < \epsilon \quad \text{for all } n > N.$

In this notation, whatever is given is written before the semi-colon. Here 'given ϵ greater than 0'.

The symbol $\exists$ (reversed E) taken together with the next following stop means 'there is (or there exists)...such that'. Here 'there exists N such that $|s_n - s| < \epsilon$ for all $n > N$.'

In this book symbolism of this kind will be used from time to time, but not on every possible occasion. It is the experience of many students of mathematics that arguments are easier to follow if brevity is not made the prime consideration and if symbolic statements are relieved by verbal sentences.

(5) You may find that a graphical representation helps to clarify the notion of limit. Referred to axes Ox, Oy, a sequence can be represented by a set of isolated points whose x-coordinates are $1, 2, \ldots, n, \ldots$, the y-coordinate of the nth point being s_n. Draw the line $y = s$ and the parallel lines $y = s - \epsilon$ and $y = s + \epsilon$; the part of the plane between these parallels forms a band of width 2ϵ.

Then the statement $s_n \to s$ means that, however small ϵ is, there is a vertical line $x = N$ such that all representative points of the sequence to the right of it lie inside the band between $y = s - \epsilon$ and $y = s + \epsilon$.

Exercises 2 (a)

Notes on these exercises are given on p. 172.

1. Define a sequence s_n satisfying the following requirements:

   (a) $0 < s_n < 1$ for all n,
   (b) there is no n for which $s_n = \frac{1}{2}$,
   (c) $s_n \to \frac{1}{2}$ as $n \to \infty$.

For each of the sequences defined in **2–6**, state whether or not it tends to a limit. If a sequence has a limit, make an (ϵ, N) table as in §2.2, taking $\epsilon = 10^{-3}$ and any other values that you like.

2. $s_n = \dfrac{3n}{n+3}$.

3. $s_n = \left(\dfrac{3n}{n+3}\right)^2$.

4. $s_n = 1/n$ if n is a prime number; $s_n = 0$ if n is not prime.

5. $s_n = \sqrt{(n+1)} - \sqrt{n}$.

6. $s_n = 1/\phi(n)$, where $\phi(n)$ is the number of integers which are factors of n (counting 1 and n as factors).

7. Prove that, if $s_n \to 0$ and $|t_n| < |s_n|$ for all n, then $t_n \to 0$.

8. Prove that a sequence cannot tend to more than one limit (as has been tacitly assumed in §2.3). i.e. prove that, if $s_n \to s$ and $s_n \to s'$, then $s = s'$.

2.4. Sequences tending to infinity

We need a concise description of the behaviour of a sequence like

$$1, \sqrt{2}, \sqrt{3}, \ldots, \sqrt{n}, \ldots,$$

the members of which exceed any assigned number for all large enough values of n.

Definition. *The sequence s_n is said to* tend to infinity *if, given A (however large), there exists N such that*

$$s_n > A \quad \text{for all } n > N.$$

We use the arrow notation and write

$$s_n \to \infty.$$

We must emphasise the difference between $s_n \to s$ and $s_n \to \infty$. In the former, s is a number and we can if we wish measure the closeness of s_n to s by the smallness of $s_n - s$. *Infinity* (∞) *is not a number* and the word 'infinity' has not yet any meaning in this book except when it follows the words 'tends to'. Any attempted manipulation of the symbol ∞ such as $s_n - \infty$ would be nonsense. The reader is reminded that the adjective *infinite* was used in §2.1 meaning 'unending' and not in any sense as measuring magnitude.

We have explained $s_n \to \infty$. The phrase 'n tends to infinity' and the corresponding statement in symbols '$n \to \infty$' likewise express the unending growth of n which is thought of in all the definitions we have laid down, to cover in turn

$$s_n \to 0, \quad s_n \to s, \quad s_n \to \infty.$$

After all this, you should not need to be warned again that '$n = \infty$' is nonsense.

The more explicit notation $$\lim_{n \to \infty} s_n = s$$

is sometimes used instead of lim $s_n = s$. For an illustration of its utility see exercise 2 (*d*), 10 (p. 34) in which there are two variables n and x.

We go on to consider other possible modes of behaviour of a sequence s_n as $n \to \infty$.

Suppose that $$s_n = 1000 - 2^n.$$

Here s_n is negative as soon as n is greater than 9 and, if n is large enough, s_n can be made numerically greater (algebraically less!) than any assigned number. The following definition is appropriate.

Definition. $s_n \to -\infty$ *as $n \to \infty$ if*

$$A; \; \exists \, N. \; s_n < -A \quad \text{for all } n > N.$$

In situations in which it is necessary to stress the distinction between $s_n \to \infty$ and $s_n \to -\infty$, the former may be written $s_n \to +\infty$.

Sequences such as $s_n = (-1)^n$ or $s_n = (-1)^n n$ do not tend to a limit or to $+\infty$ or to $-\infty$. It is convenient (but not vital) to have a name for such sequences.

Definition. *If s_n does not tend to a limit or to $+\infty$ or to $-\infty$, we say that s_n oscillates (or is an oscillating sequence). If s_n oscillates and is bounded, it oscillates finitely. If s_n oscillates and is not bounded, it oscillates infinitely.*

N.B. $s_n = (-1)^n/n$ is *not* an oscillating sequence.

Exercises 2 (*b*)

Notes on these exercises are given on p. 172.

For each sequence s_n defined in **1–6**, state whether it tends to a limit (finite or infinite) or oscillates.

1. $100n^{-1} + (-1)^n$, $100 + (-1)^n n^{-1}$, $100 + (-1)^n n$.

2. $a + b(-1)^n$, where a and b are constants.

3. $n^2\{1 + (-1)^n\}$, $n^2 + (-1)^n n$, $an^2 + b(-1)^n n$.

4. The remainder when n is divided by 3.

5. $\frac{1}{2} + \frac{1}{4} + \frac{1}{8} + \ldots + (\frac{1}{2})^n$.

6. $(1 + 2 + 3 + \ldots + n)/n^2$, $\{1 - 2 + 3 - 4 + \ldots - (-1)^n n\}/n$.

7. Give a value of N such that, if $n > N$, $n^2 - 4n > 10^6$.

Establish the truth or falsity of the statements in each of **8–10**. This means that if a statement is true you have to prove it; if it is false, construct a *counter-example*, i.e. an example satisfying the hypothesis but not the conclusion.

8. If $s_{n+1} - s_n$ oscillates finitely, then s_n oscillates.

9. If $s_{n+1} - s_n$ oscillates infinitely, then s_n oscillates infinitely.

10. If, given K (however large), we can find N for which $s_N > K$, then $s_n \to \infty$.

2.5. Sum and product of sequences

The theorems of this section are of every-day use in questions such as the following.

Example. How does
$$\frac{n^2+4n-3}{2n^2+3n+5}$$

behave as $n \to \infty$?

We think intuitively that the multiples of n and the constant terms will be negligible for large n in comparison with the terms in n^2, and so the sequence will have the same limit as $n^2/2n^2$, that is to say $\frac{1}{2}$. A less rough argument would be to start by writing
$$s_n = \frac{1+(4/n)-(3/n^2)}{2+(3/n)+(5/n^2)}.$$

If we then assume that the limit of the sum of two or more sequences is the sum of the separate limits of the sequences we can say
$$\lim \left(1+\frac{4}{n}-\frac{3}{n^2}\right) = 1+\lim\frac{4}{n}-\lim\frac{3}{n^2} = 1+0+0 = 1$$

and similarly the limit of the denominator of s_n is 2.

If we further assume that the limit of the quotient of two sequences is the quotient of their limits we have
$$\lim s_n = \tfrac{1}{2}.$$

We shall now give formal statements and proofs of theorems such as have been used in this example.

Theorem 2.51. *If s_n and t_n are null sequences so is s_n+t_n.*

The truth of this is patent. On analysing to ourselves why this is so, we argue that by taking n large enough we can make s_n arbitrarily small and also t_n arbitrarily small, and this implies the smallness of s_n+t_n. We have to express this formally.

Proof. Given ϵ, we can find N_1 such that
$$-\epsilon < s_n < \epsilon \quad \text{for all } n > N_1.$$

We can find N_2 such that
$$-\epsilon < t_n < \epsilon \quad \text{for all } n > N_2.$$

If N is the greater of N_1 and N_2, i.e. in the notation of §1.7, $N = \max(N_1, N_2)$, then, for $n > N$, both the above sets of inequalities hold and we have, adding them,
$$-2\epsilon < s_n+t_n < 2\epsilon.$$

(Observe now that, when ϵ is allowed to take any positive value, the 2ϵ serves just as well as ϵ for 'any positive number, as small as you like'.) Therefore $s_n + t_n$ is a null sequence, and the theorem is proved.

Theorem 2.52. *If s_n is a null sequence and t_n is a bounded sequence, then $s_n t_n$ is a null sequence.*

Proof.
$$\exists K. \quad |t_n| < K \quad \text{for all } n.$$

Also, $\qquad \epsilon > 0; \ \exists N. \ |s_n| < \epsilon \quad \text{for all } n > N.$

Therefore $\qquad |s_n t_n| < K\epsilon \quad \text{for all } n > N.$

Therefore $s_n t_n$ is a null sequence. (Note the remark in brackets at the end of the last proof.)

Corollary. *If s_n is a null sequence and c a constant, then cs_n is a null sequence.*

We shall now step from null sequences to general sequences.

Theorem 2.53. *If $s_n \to s$ and $t_n \to t$, then*

$$\text{(i)} \quad s_n + t_n \to s + t,$$

$$\text{(ii)} \quad s_n t_n \to st.$$

Proof. (i) $s_n - s$ and $t_n - t$ are null sequences; therefore so is their sum $s_n + t_n - (s + t)$. This proves (i).

(ii) $s_n t_n - st = (s_n - s) t_n + s(t_n - t).$

In the first term on the right-hand side, $s_n - s$ is a null sequence and t_n is bounded; therefore their product is a null sequence. The second term, being the null sequence $t_n - t$ multiplied by the constant s is a null sequence. So the right-hand side, being the sum of two null sequences is a null sequence and therefore

$$s_n t_n \to st.$$

Theorem 2.54. *If $s_n \to s$ and $t_n \to t$, where $t \neq 0$, then*

$$\frac{s_n}{t_n} \to \frac{s}{t}.$$

Proof. We shall prove that

$$\frac{1}{t_n} \to \frac{1}{t}$$

and theorem 2.54 will then follow from theorem 2.53 applied to the product of the two sequences s_n and $1/t_n$.

We wish therefore to prove that

$$\frac{1}{t_n} - \frac{1}{t}, \quad \text{that is to say} \quad \frac{t - t_n}{t_n t},$$

is a null sequence.

We can choose N such that, for all $n > N$,

$$|t_n| > \tfrac{1}{2}|t|$$

and so

$$\frac{1}{|t_n t|} < \frac{2}{|t|^2}.$$

Then $t - t_n$ is a null sequence and $1/t_n t$ is a bounded sequence and theorem 2.52 shows that $(t - t_n)/t_n t$ is a null sequence. **❙**

Exercises 2 (c)

Notes on these exercises are given on p. 172.

1. Discuss the behaviour as $n \to \infty$ of the sequences whose nth terms are

$$\left(\frac{n-3}{3n+1}\right)^3, \quad \left(\frac{3n+1}{n-3}\right)^3, \quad \frac{n(n-1)}{(n-2)(n-3)(n-4)}.$$

2. Discuss the behaviour as $n \to \infty$ of the general rational function of n

$$R(n) = \frac{a_0 n^p + a_1 n^{p-1} + \ldots + a_p}{b_0 n^q + b_1 n^{q-1} + \ldots + b_q},$$

where p and q are positive integers.

2.6. Increasing sequences

Definition. *If $s_{n+1} \geqslant s_n$ for all values of n we call s_n increasing.*

It is useful to regard increase in the wide sense, allowing the possibility of equality at any of the steps from n to $n+1$. If $s_{n+1} > s_n$ for all n, we call s_n *strictly increasing*.

If $s_{n+1} \leqslant s_n$ for all values of n, we call s_n *decreasing*. A word which usefully covers either increasing or decreasing is *monotonic*.

Examples. Which of the following sequences are increasing or decreasing?

(1) $\dfrac{n}{n^2+1}$,   (2) $\dfrac{n^2+1}{n}$,   (3) 1,   (4) $n+(-1)^n$,   (5) $2n+(-1)^n$.

We shall prove that a monotonic sequence has the very important property that it must tend to a limit or to $+\infty$ or $-\infty$. In other words, a monotonic sequence cannot oscillate.

Theorem 2.6. *An increasing sequence tends either to a limit or to $+\infty$.*

Proof. Let s_n be the sequence. There are two possibilities. Either

(1) a number A can be found such that $s_n \leqslant A$ for all n; or

(2) whatever number A is taken, there is a value of N for which $s_N > A$.

Let us deal first with the possibility (2). Since s_n is increasing, $s_n > A$ not merely for $n = N$ but for $n \geqslant N$. Then, straight from the definition, $s_n \to \infty$.

Take now the case (1). The number A is an upper bound of the s_n. By theorem 1.8, there is a number $s = \sup s_n$ with the properties

$$s_n \leqslant s \quad \text{for all } n$$

and $\qquad s_n > s - \epsilon \quad$ for some particular value of n.

Since s_n is increasing, the second inequality will hold also for all values of n beyond that particular value. The two inequalities show that $s_n \to s$. The theorem is proved.

The most useful case of the theorem may be summed up in the statement

A bounded increasing sequence tends to a limit.

The corresponding theorem that a decreasing sequence tends to a limit or to $-\infty$ can be proved, either by an analogous argument or by using the fact that if s_n decreases then $-s_n$ increases.

2.7. An important sequence a^n

Let $s_n = a^n$, where a is a constant. The behaviour of the sequence as $n \to \infty$ depends on the value of a.

(1) If $a = 1$, $s_n = 1$ for all n and $\lim s_n = 1$. If $a = 0$, $\lim s_n = 0$.

(2) Suppose $a > 1$. Let $a = 1 + k$, where $k > 0$.

Then $\qquad\qquad s_n = (1 + k)^n > 1 + nk$

(by taking only the first two terms of the binomial expansion).

As $n \to \infty$, $1+nk \to \infty$ and therefore $s_n \to \infty$.

(3) Suppose $0 < a < 1$. Let $a^{-1} = 1+l$, where $l > 0$.

Then
$$0 < s_n = \frac{1}{(1+l)^n} < \frac{1}{1+nl}.$$

As $n \to \infty$, $1/(1+nl) \to 0$ and therefore $s_n \to 0$.

(4) Suppose a to be negative. If $-1 < a < 0$ and $a = -b$, so that $0 < b < 1$, it follows from (3) that $b^n \to 0$ and hence $s_n = (-b)^n \to 0$.

If $a = -1$, s_n takes the values -1 and 1 alternately and oscillates finitely.

If $a < -1$ and $a = -b$, $b > 1$, then from (2) we have $b^n \to \infty$. So $s_n = (-b)^n$ takes values, alternately negative and positive, numerically greater than any assigned number. That is to say, s_n oscillates infinitely. Summing up, we have

$$a^n \to \infty \quad (a > 1),$$

$$a^n \to 1 \quad (a = 1),$$

$$a^n \to 0 \quad (-1 < a < 1),$$

$$a^n \text{ oscillates finitely} \quad (a = -1),$$

$$a^n \text{ oscillates infinitely} \quad (a < -1).$$

Exercises 2 (d)

Notes on these exercises are given on p. 172.

1. If $s_1 > 0$ and $s_{n+1} \geqslant Ks_n$, where $K > 1$, for all values of n, then $s_n \to +\infty$.

2. If, for all values of n, $|s_{n+1}| \leqslant K|s_n|$, where $0 < K < 1$, then $s_n \to 0$. The conclusion remains true if the hypothesis is satisfied only for $n > N$.

3. If
$$\lim \frac{s_{n+1}}{s_n} = l, \quad -1 < l < 1,$$
prove that $s_n \to 0$.

4. Discuss the behaviour, as $n \to \infty$, of the sequence a^n/n^k, where k is a positive integer.

5. Prove that, if $a > 0$, then $\sqrt[n]{a} \to 1$ as $n \to \infty$.

6. Prove that $\{1+(1/n)\}^n$ increases as n increases and that it tends to a limit. (This limit is the very important number e—see chapter 6.)

7. Prove that $\sqrt[n]{n}$ decreases as n increases, and deduce that it tends to the limit 1.

8. Give examples of sequences s_n for which

$$\lim \frac{s_{n+1}}{s_n} = 1$$

and (a) $s_n \to \infty$, (b) $s_n \to 3$, (c) $s_n \to 0$.

9. If $R(n)$ has the meaning assigned to it in exercise 2 (c), 2, prove that, if $-1 < x < 1$,
$$\lim R(n) x^n = 0.$$

Discuss the behaviour as $n \to \infty$ of $R(n) x^n$ for other values of x.

10. Prove that, if $x \neq -1$, then
$$u_n(x) = \frac{x^n - 1}{x^n + 1}$$

tends to a limit as $n \to \infty$. Show in a diagram the graph of

$$y = \lim_{n \to \infty} u_n(x).$$

11. Prove that, if $-1 < x < 1$,

$$u_n = \frac{m(m-1) \dots (m-n+1)}{n!} x^n = \binom{m}{n} x^n$$

tends to zero as $n \to \infty$.

12. Investigate whether the following sequences have limits as $n \to \infty$ and, if so, give the limits.

(i) $\sqrt[n]{(3^n + 2^n)}$, and extend to numbers other than 3, 2;

(ii) $\dfrac{a^n - b^n}{a^n + b^n}$, where $a > 0$ and $b > 0$;

(iii) $\dfrac{n}{n^2 + 1} + \dfrac{n}{n^2 + 2} + \dots + \dfrac{n}{n^2 + n}$;

(iv) $\sqrt[n]{(n!)}$;

(v) $n - \sqrt{(n+a)(n+b)}$;

(vi) $2^{n^2}/n!$.

2.8. Recurrence relations

Hitherto we have supposed s_n to be given by an explicit formula in terms of n. In practice a sequence is often determined by a relation connecting two or more successive members of it, and it may or may not be possible to 'solve' this for s_n. We illustrate by examples of useful types.

Example 1. *The linear recurrence relation (or difference equation) with constant coefficients.* Suppose that we are given that

$$s_n = s_{n-1} + s_{n-2} \quad (n \geqslant 2),$$

$$s_0 = 1, \quad s_1 = 1.$$

(This is one of the cases in which $n = 0$ is a more appropriate starting value than $n = 1$, because it is useful—as is shown in books on algebra—to associate with the sequence a *generating function* $s_0 + s_1 x + s_2 x^2 + \dots$.)

The data enable us to write down as many terms of the sequence as we please

$$1, \ 1, \ 2, \ 3, \ 5, \ 8, \ 13, \ \dots.$$

We shall obtain an explicit formula for s_n. To do so, substitute the trial solution

$$s_n = A\alpha^n + B\beta^n,$$

where A, B, α, β are constants. We see that $s_n = s_{n-1} + s_{n-2}$ if α and β are the roots of the equation

$$t^2 - t - 1 = 0.$$

Knowing α, β, we can then determine A and B from the given values $s_0 = 1$, $s_1 = 1$. We find

$$\alpha = \tfrac{1}{2}(1 + \sqrt{5}), \quad \beta = \tfrac{1}{2}(1 - \sqrt{5}).$$

$$A + B = 1, \quad A\alpha + B\beta = 1$$

and hence

$$s_n = \frac{1}{\sqrt{5}} \left\{ \left(\frac{1 + \sqrt{5}}{2} \right)^{n+1} - \left(\frac{1 - \sqrt{5}}{2} \right)^{n+1} \right\}.$$

Example 2. Investigate the sequence defined by

$$s_{n+1} = \sqrt{(s_n + a)}, \quad s_1 = b,$$

where a and b are given positive numbers.

Remarks. (i) There is no way of obtaining a compact formula for s_n in terms of n.

(ii) If we provisionally *assume* that s_n tends to a limit s, we can say what the limit must be. For let $n \to \infty$ and we have

$$s = \sqrt{(s + a)}.$$

Since every s_n is positive, s is the positive root of the quadratic $s^2 = s+a$.

(iii) Sequences obeying a simple recurrence relation are very commonly monotonic. See whether $s_{n+1}-s_n$ has a fixed sign.

Solution of example 2. Let α be the positive root of the quadratic
$$s^2 = s+a.$$

We have $\qquad\qquad s_{n+1}^2 - s_n^2 = a + s_n - s_n^2.$

If $s_n > \alpha$, the right-hand side is negative and so
$$s_{n+1} < s_n.$$

Also $\qquad\qquad s_{n+1}^2 - \alpha^2 = (s_n+a) - (\alpha+a)$
$$= s_n - \alpha.$$

This shows that, if $s_n > \alpha$, then $s_{n+1} > \alpha$.

Suppose that $b = s_1 > \alpha$. Then, by induction, s_n forms a decreasing sequence with $s_n > \alpha$ for all n.

By theorem 2.6, s_n tends to a limit s where $s \geqslant \alpha$, and, by the argument of remark (ii), $s = \alpha$.

Similarly if $s_1 < \alpha$, then s_n forms an increasing sequence with limit α. If $s_1 = \alpha$, every $s_n = \alpha$.

Graphical representation. For recurrence relations of the special form $s_{n+1} = f(s_n)$ the movements of s_n may be traced by drawing the graphs of the curve C and the line L whose respective equations are $y = f(x)$ and $y = x$ and making the following construction. (See fig. 1.)

Let P_1 be the point on C whose x-coordinate is s_1. Its y-coordinate is therefore $f(s_1) = s_2$. Through P_1 draw a horizontal line to meet L in Q_1, then through Q_1 a vertical line to meet C in P_2. P_2 has coordinates (s_2, s_3). We continue from P_2 as we did from P_1.

Exercises 2 (*e*)
Notes on these exercises are given on p. 173.

Investigate the behaviour as $n \to \infty$ of s_n if s_n is given by the recurrence relations stated in **1–6**.

1. $3s_n = 2s_{n-1} + s_{n-2}$ $\quad(s_0 = 7, s_1 = 3)$.

2. $s_{n+1} = 2/(1+s_n)$ $\quad(s_1 = 0)$.

3. $s_{n+1} = s_n^2 + k$, where $0 < k < \frac{1}{4}$ and s_1 lies between the roots of the equation $x^2 - x + k = 0$.

4. $s_{n+1} = \dfrac{2(2s_n + 1)}{s_n + 3}$ $(s_1 = 3)$.

5. $s_{n+1} = \dfrac{5s_n - 3}{s_n + 1}$ $(s_1 = 2)$.

Consider also other values of s_1 (e.g. $s_1 = \frac{1}{5}$). What happens if $s_1 = \frac{5}{7}$?

6. $s_n = (s_{n-1}^2 + s_{n-2} + 2)^{1/3}$ $(s_1 = 1, s_2 = \frac{3}{2})$.

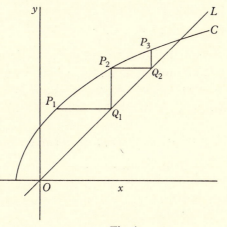

Fig. 1

7. Prove that, if
$$u_1 = 3 \quad \text{and} \quad u_{n+1} = (u_n + 5)^{1/2},$$
then u_n tends to a limit l, and give l to two decimal places.

Prove that
$$0 < u_{2n+1} - l < 30^{-n}(u_1 - l).$$

Note. In numerical applications of analysis, it is vital to know with what rapidity we can rely on a sequence approaching its limit. This information is given by the last inequality.

8. Discuss the possible limits of the sequence defined by
$$s_{n+1} = \frac{6s_n^2 + 6}{s_n^2 + 11}.$$
Prove that, if $s_n > 3$, then

(i) $3 < s_{n+1} < s_n$,   (ii) $s_{n+1} - 3 < \frac{9}{10}(s_n - 3)$.

9. Prove that, if
$$a_1 > b_1 > 0 \quad \text{and} \quad a_{n+1} = \tfrac{1}{2}(a_n + b_n), \quad b_{n+1} = 2a_n b_n/(a_n + b_n),$$
then $a_n > a_{n+1} > b_{n+1} > b_n$. Prove that, as $n \to \infty$, a_n and b_n both tend to the limit $\sqrt{(a_1 b_1)}$.

2.9. Infinite series

It is likely that you have learned from books on algebra that, if the common ratio of a geometric progression is numerically less than 1, the progression has a 'sum to infinity'. The simplest illustration of this is

$$\tfrac{1}{2}+\tfrac{1}{4}+\tfrac{1}{8}+\tfrac{1}{16}+\dots.$$

Here the sum s_n of the first n terms is

$$s_n = 1-2^{-n}.$$

As $n \to \infty$, $s_n \to 1$ and the sum to infinity is 1. So there is no new idea in the summation of an infinite series; we have only to examine the behaviour, as $n \to \infty$, of the sequence of numbers s_n, where s_n is the sum of the first n terms. We shall state this formally.

Let u_n be defined for all positive integral values of n. Define

$$s_n = u_1+u_2+\dots+u_n$$

or, more shortly, $s_n = \sum\limits_{r=1}^{n} u_r$.

If, as $n \to \infty$, s_n tends to a finite limit s, we say that the infinite series

$$u_1+u_2+u_3+\dots$$

or $\sum\limits_{n=1}^{\infty} u_n$ converges (or, is convergent) and that s is its sum.

The number u_n is the nth *term* of the infinite series and s_n is the sum of the first n terms. (It may sometimes be more appropriate to start with a term with zero suffix u_0, for example, in the series $a_0+a_1x+a_2x^2+\dots$; s_n will then denote the sum of $n+1$ terms.)

You should note carefully that, when applied to infinite series, the meaning of the word *sum* has been widened from its use in algebra. Hitherto it has meant the number which is got by adding the numbers contained in some given finite set. Now it can be the limit of a sequence. The main reason for this cautionary remark is to guard ourselves against tacitly assuming that properties of a sum (in the restricted sense) carry over into properties of a sum (in the general sense). For instance, if $a+b+c = s$, then by the associative law of algebra,

$$ka + kb + kc = ks.$$

The corresponding property is in fact true of the sum of an infinite series, i.e. if $u_1 + u_2 + u_3 + \dots$ converges to sum s, then $ku_1 + ku_2 + ku_3 + \dots$ converges to sum ks. But it is a property of limits and must be proved as such.

An infinite series which is not convergent is called *divergent*. For example, each of the series

$$1 + 1 + 1 + \dots$$

and
$$1 - 1 + 1 - \dots$$

is divergent. From §2.4 we see that for a divergent series there are four different possible ways in which s_n may behave; it may tend to $+\infty$ or to $-\infty$ or may oscillate finitely or infinitely. If a series does not converge, it is generally enough just to label it as divergent, but it is sometimes easy and convenient to incorporate further information. For instance, we could say that the series $\Sigma(-1)^n$ oscillates and the series

$$\Sigma\{-1 + (-1)^n\}$$

diverges to $-\infty$.

The words *converge* and *diverge* are commonly applied to sequences as well as to series, e.g. if $s_n = 2 - (\frac{1}{2})^n$, then s_n converges to the limit 2. If s_n does not tend to a finite limit we call it divergent; a more specific statement may be made such as that $s_n = -2^n$ diverges to $-\infty$.

Before we discuss general properties of infinite series it will help the reader to be thoroughly familiar with some particular series of simple types which will illustrate the later theorems. We shall take two important series, the first being the geometric series.

2.10. The geometric series Σx^n

Theorem 2.10. *The infinite series*

$$1 + x + x^2 + \dots + x^n + \dots$$

converges if and only if $-1 < x < 1$.

Proof. Here $u_n = x^{n-1}$ and

$$s_n = \begin{cases} \dfrac{1-x^n}{1-x} & (x \neq 1), \\ n & (x = 1). \end{cases}$$

If $x = 1$, $s_n \to \infty$ as $n \to \infty$. If $x \neq 1$,

$$s_n = \frac{1}{1-x} - \frac{x^n}{1-x}.$$

From §2.7, $x^n \to 0$ if $-1 < x < 1$ and the series converges if $-1 < x < 1$, its sum being $1/(1-x)$.

If $x > 1$, $x^n \to \infty$ and $s_n \to \infty$.

If $x = -1$, $s_n = 1$ if n is odd and 0 if n is even and so oscillates finitely.

If $x < -1$, s_n oscillates infinitely.

Hence the series converges *only if* $-1 < x < 1$.

2.11. The series Σn^{-k}

Observe that the meaning of a power n^k is determined by the index laws of algebra only if k is rational. The extension to irrational k is best deferred until chapter 6 (p. 108). Meanwhile we shall assume that the indices of any powers with which we deal are rational. The theorems concerned (such as theorem 2.11) remain true whether the indices are rational or irrational.

Theorem 2.11. *The infinite series*

$$\frac{1}{1^k} + \frac{1}{2^k} + \frac{1}{3^k} + \ldots + \frac{1}{n^k} + \ldots$$

(*where k is a constant*) *converges if $k > 1$ and diverges if $k \leqslant 1$.*

Remarks. If $k = 1$, the terms are in 'harmonic progression' and the series $\Sigma(1/n)$ is often called the harmonic series.

The geometric series of the theorem 2.10 was easy to deal with for the reason that we could find a simple explicit formula for s_n in terms of n. This is not possible for the series $\Sigma(1/n^k)$ even for simple values of k like 1 or 2. In practice, given an infinite series, it is most unlikely that a simple expression can be found for the sum of n terms. Some sort of approximation is nearly always necessary. The reader will see in a moment the device by which it can be carried out for this particular series. More general methods will be developed in chapter 5.

We shall prove theorem 2.11 first for the 'border-line' value $k = 1$, which is the greatest value of k giving divergence. This, being the case of the theorem in which the issue between convergence and divergence is most finely balanced, might be expected to be the most difficult. In fact, on account of the simplicity of detail, it is as easy as the other cases.

Proof of theorem. We wish to prove that the series

$$1 + \tfrac{1}{2} + \tfrac{1}{3} + \tfrac{1}{4} + \tfrac{1}{5} + \tfrac{1}{6} + \tfrac{1}{7} + \tfrac{1}{8} + \dots$$

diverges. This will be done if, by taking enough terms, we can make their sum larger than any assigned number. The following inequalities use the fact that the terms decrease

$$\tfrac{1}{3} + \tfrac{1}{4} > \quad 2 \times \tfrac{1}{4} = \tfrac{1}{2},$$

$$\tfrac{1}{5} + \tfrac{1}{6} + \tfrac{1}{7} + \tfrac{1}{8} > \quad 4 \times \tfrac{1}{8} = \tfrac{1}{2},$$

$$\tfrac{1}{9} + \tfrac{1}{10} + \dots + \tfrac{1}{15} + \tfrac{1}{16} > 8 \times \tfrac{1}{16} = \tfrac{1}{2}$$

and so on. The sum of the terms in each of the successive blocks of 2, 4, 8, 16, ... terms is greater than $\tfrac{1}{2}$. Taking in now the first and second terms 1 and $\tfrac{1}{2}$, we have proved that the sum of the first $2 + 2 + 4 + 8 + \dots + 2^{m-1}$ (i.e. 2^m) terms of the series is greater than $1 + \tfrac{1}{2}m$. Therefore the series $\Sigma(1/n)$ diverges to $+\infty$.

Suppose now that $k < 1$. Then $1/n^k > 1/n$, and so the sum of any set of terms of $\Sigma(1/n^k)$ is greater than the sum of the corresponding terms of $\Sigma(1/n)$. This latter series has just been proved to diverge to $+\infty$. Therefore so does $\Sigma(1/n^k)$ for $k < 1$.

Take now $k > 1$. We have to prove convergence. The same device, of grouping terms in blocks of 2, 4, 8, ... succeeds, but we must now find approximations which are *greater* than the sums of the blocks. We have

$$\frac{1}{2^k} + \frac{1}{3^k} < \frac{2}{2^k} = \frac{1}{2^{k-1}},$$

$$\frac{1}{4^k} + \frac{1}{5^k} + \frac{1}{6^k} + \frac{1}{7^k} < \frac{4}{4^k} = \frac{1}{4^{k-1}},$$

$$\frac{1}{8^k} + \frac{1}{9^k} + \dots + \frac{1}{15^k} < \frac{8}{8^k} = \frac{1}{8^{k-1}}.$$

Now the series

$$1 + \frac{1}{2^{k-1}} + \frac{1}{4^{k-1}} + \frac{1}{8^{k-1}} + \dots$$

is a convergent geometric series, with sum to infinity t say. Then s_N, the sum of the first N terms of $\Sigma(1/n^k)$ increases with N and is always less than t. So, by theorem 2.6, s_N tends to a finite limit as $N \to \infty$. That is to say, the series converges. $\blacksquare$

Note on the case $k < 1$. Instead of deducing this from the case $k = 1$, we could argue directly

$$1 + \frac{1}{2^k} + \dots + \frac{1}{n^k} > n\left(\frac{1}{n^k}\right)$$

and $n^{1-k} \to \infty$.

Numerical illustration. Find a value of n large enough to make the sum of n terms of

$$1 + \tfrac{1}{2} + \tfrac{1}{3} + \dots$$

greater than 20.

The reader should not disdain numerical work, which keeps him in touch with reality. We will work this example for him. The proof showed that the sum of the first 2^m terms is greater than $1 + \tfrac{1}{2}m$. So 2^{38} terms will certainly be enough. Now $2^{38} > 10^{11}$ and a twelve-figure number in this context is surprisingly large. We naturally sacrificed accuracy in approximating to the sums of blocks of terms. But more refined methods would show that the number of terms which must be taken to give a sum greater than 20 exceeds 10^8. These calculations show, in homely language, that *the divergence of this series is slow.*

Exercises 2 (*f*)

Notes on these exercises are given on p. 173.

1. For the series $\Sigma 1/n^2$ estimate a value of N which will ensure that the sum of all terms after the Nth is less than 10^{-4}.

Make the same calculation for the geometric series $\Sigma(0.99)^n$. What do you conclude about the relative 'slowness—or quickness—of convergence' of the two series?

2. Find for what values of r the following series converge, and sum them

$$r + \frac{r}{1+r} + \frac{r}{(1+r)^2} + \frac{r}{(1+r)^3} + \dots,$$

$$r^2 - \frac{r^2}{1+r^2} + \frac{r^2}{(1+r^2)^2} - \frac{r^2}{(1+r^2)^3} + \dots,$$

$$1 + 2r + 3r^2 + 4r^3 + \dots.$$

3. Estimate the sum $\sum\limits_{n=0}^{60} (\tfrac{2}{3})^n$.

4. Prove that, if $a > 0$ and $b > 0$, the series whose nth term is $1/(a+nb)$ diverges.

5. Sum to m terms and to infinity the series whose nth term is

$$\frac{1}{n(n+1)(n+2)}.$$

Deduce the convergence of Σn^{-3}. Extend this to a denominator which is a product of k instead of three consecutive integers.

2.12. Properties of infinite series

The following properties of infinite series are, for the most part, immediate adaptations of results about limits of sequences, and proofs can be supplied by the reader.

(1) *The convergence or divergence of a series is unaffected if a finite number of terms are inserted, or suppressed, or altered.*

Illustration. If $\sum\limits_{1}^{\infty} u_n$ converges to sum s, then $\sum\limits_{k+1}^{\infty}$ converges to sum $s - u_1 - \ldots - u_k$. If the former series diverges, so does the latter.

(2) *If $u_1 + u_2 + \ldots$ converges to sum s and $v_1 + v_2 + \ldots$ converges to sum t, then the series*

$$(u_1 + v_1) + (u_2 + v_2) + \ldots$$

converges to sum $s + t$.

Exercise. Prove, more generally, that the series whose nth term is $au_n + bv_n$, where a and b are constants, converges to sum $as + bt$.

(3) *If $u_1 + u_2 + \ldots$ converges, then $\lim u_n = 0$.*

Proof. $u_n = s_n - s_{n-1}$. Both s_n and s_{n-1} tend to the same limit s. Therefore, by theorem 2.53 (i), $\lim u_n$ exists and $\lim u_n = s - s = 0$. ∎

Note carefully that the converse of (3) is false. The example $u_n = 1/n$ shows that it is possible to have $\lim u_n = 0$ and Σu_n divergent. In other words (see exercise 2 (g), 1)

The condition $\lim u_n = 0$ is necessary but not sufficient for the convergence of Σu_n.

(4) *If Σu_n is convergent, then so is any series whose terms are obtained by bracketing the terms of Σu_n in any manner, and the two series have the same sum.*

It is thus legitimate to insert brackets in a convergent series, but not to omit them. The series $1-1+1-1+\ldots$ oscillates, but the series

$$(1-1)+(1-1)+\ldots$$

converges to sum 0.

(5) *If $u_n \geqslant 0$ for every n, then Σu_n either converges or diverges to $+\infty$.*

A necessary and sufficient condition for convergence is that there exists K such that $\sum_1^N u_n < K$ for all N; and then $\sum_1^\infty u_n \leqslant K$.

This is a restatement of theorem 2.6.

(6) *If, for every n,*

    (i) $u_n \geqslant 0, v_n \geqslant 0$,

    (ii) $u_n \leqslant Kv_n$, *where K is constant,*

    (iii) Σv_n *converges,*

then Σu_n converges. Also $\Sigma u_n \leqslant K\Sigma v_n$.

The reader should deduce this from (5) and should formulate a corresponding result for divergence.

We shall take up in chapter 5 the more systematic investigation of infinite series.

Exercises 2 (g)

Notes on these exercises are given on pp. 173–4.

1. (N.B.) Explain *necessary condition, sufficient condition*. Which of the following are (*a*) necessary, (*b*) sufficient, for the real numbers p, q to be equal?

    (i) $p^2+q^2 = 2pq$,   (ii) $p^2 = q^2$,   (iii) $p+\dfrac{1}{q} = q+\dfrac{1}{p}$.

2. Say of each of the following conditions whether it is (*a*) necessary, (*b*) sufficient.

    (i) A * condition that the year n is a leap year is that n is a multiple of 4.

    (ii) A * condition that $pq = 0$ is that $p = 0$ and $q = 0$.

    (iii) A * condition that the corresponding angles of two triangles *ABC, DEF* are equal is that

$$\frac{BC}{EF} = \frac{CA}{FD} = \frac{AB}{DE}.$$

3. Say of each of the following series whether it converges or diverges

$$\frac{1}{3}+\frac{2}{5}+\frac{3}{7}+\ldots+\frac{n}{2n+1}+\ldots,$$

$$\frac{1}{\sqrt{2}}+\frac{1}{2}+\frac{1}{2\sqrt{2}}+\ldots+\frac{1}{2^{n/2}}+\ldots,$$

$$1-1+\tfrac{1}{2}-\tfrac{1}{2}+\tfrac{1}{3}-\tfrac{1}{3}+\tfrac{1}{4}-\tfrac{1}{4}+\ldots,$$

$$1+\frac{1}{1!}+\frac{1}{2!}+\frac{1}{3!}+\ldots.$$

4. Prove that, if $0 \leqslant a_n \leqslant 1$, then the series $a+a_1x+\ldots+a_nx^n+\ldots$ converges for $0 \leqslant x < 1$.

Extend the result, assuming that $|a_n| \leqslant k$.

5. Sum the finite series

$$\sum_{r=1}^{n} \frac{1}{r(r+2)(r+4)}, \quad \sum_{r=1}^{n} \frac{ar+b}{r(r+1)(r+2)}.$$

Deduce the sums to infinity.

6. (N.B.) If

$$s_n = \left(1+\frac{1}{n}\right)^n \quad \text{and} \quad t_n = \left(1-\frac{1}{n}\right)^{-n}$$

prove that s_n is an increasing and t_n a decreasing sequence. Prove further that, as $n \to \infty$, $t_n - s_n \to 0$ and that s_n and t_n tend to the same limit.

7. If s_n and t_n oscillate finitely, in what ways is it possible for s_n+t_n to behave (in regard to limits or oscillation)? Give an example of each. Answer the same question for $s_n t_n$.

8. Simplify the product

$$\frac{2^3-1}{2^3+1}\frac{3^3-1}{3^3+1}\frac{4^3-1}{4^3+1}\cdots\frac{n^3-1}{n^3+1},$$

and prove that it tends to a limit (to be found) as $n \to \infty$.

9. It is given that $u_{n+1} = \frac{1}{2}(u_n+A^2/u_n)$, where $n = 1, 2, 3 \ldots$, and $0 < A \leqslant u_1$. Prove that

(i) $u_{n+1} \geqslant A$ and $u_{n+1} \leqslant u_n$;

(ii) $d_{n+1} = d_n^2$, where $d_n = (u_n-A)/(u_n+A)$;

(iii) as n tends to infinity, u_n tends to A.

Taking $A^2 = 99$, $u_1 = 10$, calculate $\sqrt{11}$ correct to four places of decimals.

10. Show that, if r_0 and A are positive numbers, and

$$r_{n+1}+\frac{1}{r_n} = 2A,$$

then the condition $A \geqslant 1$ is necessary for the convergence of the sequence r_n; show that it is also sufficient in the case of $r_0 > 1$, by verifying that $r_n > 1$ for every n, and, for a suitable $c > 1$,

$$c^n |r_n-c| \leqslant |r_0-c|.$$

11. Define (if possible) an oscillating sequence satisfying

$$\frac{s_{n+1}}{s_n} \to 1.$$

Answer the same question with $-\frac{1}{2}$ in place of 1.

12. Prove that, if

$$u_n = \frac{1 \cdot 3 \cdot 5 \dots (2n-1)}{2 \cdot 4 \cdot 6 \dots 2n},$$

then nu_n^2 is an increasing sequence and $(n+\frac{1}{2})u_n^2$ is a decreasing sequence. Deduce that nu_n^2 tends to a limit.

13. Establish the truth or falsity of each of the statements (i)–(iii).

(i) If the sequence s_n increases, so does the sequence

$$(s_1 + s_2 + \dots + s_n)/n.$$

(ii) For all n, it is known that $s_n > s_{n+1}$, $t_n < t_{n+1}$ and $s_n > t_n$. Then both sequences s_n, t_n tend to limits, s, t and $s > t$.

(iii) If $s_{n+1} - s_n$ oscillates infinitely, then s_n oscillates infinitely.

14. Prove that, if s_n tends to the limit s, then

$$\frac{s_1 + s_2 + \dots + s_n}{n}$$

also tends to s.

15. Find the sum of the first $n+1$ terms of the series

$$1 + 2x + 3x^2 + \dots + (n+1)x^n + \dots.$$

Prove that, if $|x| < 1$, the series converges to the sum $1/(1-x)^2$.

3

CONTINUOUS FUNCTIONS

3.1. Functions

You are familiar with the dependence of one real number y on another real number x which is commonly determined by formulae such as

$$y = 1/(x^2+1) \quad \text{or} \quad y = \sqrt{\{(2-x)\,(x-1)\}}.$$

In the former y is defined for all values of x and in the latter only for $1 \leqslant x \leqslant 2$. In either case, if a value of x is assigned, we can calculate the value of y which corresponds to it and there is only one such value of y (if it is understood that the $\sqrt{}$ is positive). Here x may be described as the *independent variable* and y as the *dependent variable*.

These two examples are simple illustrations of functions. We go on to express in general terms the meaning of *function*.

Let X be a set of numbers x and Y a set of numbers y. If rules are given by which, to each x in X, a corresponding y in Y is assigned, these rules determine a function defined for x in X.

Notes. (1) Observe that, to a given x, there is just one corresponding y. The same y may correspond to more than one x. For instance, in the example $y = 1/(x^2+1)$, $y = \frac{1}{2}$ corresponds to both $x = 1$ and $x = -1$. In other words, the correspondence between X and Y is in general many-one and is only exceptionally one-one.

In the ordinary graphical representation, no line parallel to Oy cuts the curve in more than one point.

(2) You may come across in mathematical literature other words and phrases equivalent to those used here. A function defined in X with values in Y is often called a transformation, or a mapping, of X into Y. According to the definition, Y may include values which are not taken by y for any value of x. If every value of y in Y is taken for some x in X, we can say that the mapping of X is *onto* Y.

(3) The usual notation for a function is the letter f. If more letters are needed, g, ϕ, F are commonly used. The value y which a function f takes for a particular x in X is written $f(x)$, so that $y = f(x)$.

(4) Strictly speaking, the function f is the set of all pairs of numbers (x, y) which are related by the rules defining the correspondence. $f(x)$ is the 'function-value' for the number x. It is convenient, and does not lead to error, to speak simply of *the function $f(x)$*.

A few further illustrations will bring out the full meaning of the definition of function. In each of these, satisfy yourself that the rules are adequate to determine a unique value of y for each value of x concerned.

(1) $$y = 1 \quad \text{for} \quad 0 \leqslant x \leqslant 1,$$

$$y = 0 \quad \text{for} \quad x < 0 \quad \text{and for} \quad x > 1.$$

In physical applications the independent variable is commonly the time; this function could represent, for example, a force of unit magnitude which acts for unit time and then ceases.

(2) Suppose that $y^2 = x$. This equation assigns values to y for $x \geqslant 0$, and, if $x > 0$, there are two values of y equal in magnitude and of opposite signs. We said in the definition that a function is to be *single-valued*. So $y^2 = x$ does not define in our sense y as a function of x. We may, however, dissect it into two functions $y = +\sqrt{x}$ and $y = -\sqrt{x}$ each defined for $x \geqslant 0$.

(3) $$y = \frac{x}{(x-1)(x-2)}.$$

This function is defined for all values of x except $x = 1$ and $x = 2$.

(4) $y = $ the largest prime factor of x.

This statement has meaning only when x is an integer.

(5) $s_n = 1/n$, where n is a positive integer. Sequences are functions of a special type, in which the independent variable is restricted to be a positive integer. The same defining formula with n replaced by x, namely, $y = 1/x$, may define a function for values of x other than positive integers (here, for all values of x except $x = 0$).

(6) $$y = 0 \quad \text{when } x \text{ is irrational,}$$

$$y = 1 \quad \text{when } x \text{ is rational.}$$

An 'unnatural' function like this may seem to have little interest. Such functions are, in fact, of use in analysis in deciding for how wide a class of functions some proposed theorem is true. We shall come across illustrations of this later in the book.

(7) y is the temperature in degrees at time x at a given place.

This type of function is very common in science and in every-day life. It differs from the examples (1)–(6) in that there is no analytical formula by which it can be represented. In practice the function-values may be given by a graph (drawn, say, by a pen attached to a recording thermometer). These values are then known within the limits of accuracy which can be attained in observation.

3.2. Behaviour of $f(x)$ for large values of x

We described in chapter 2 the various ways in which a sequence s_n may behave as the variable n tends to infinity. The same descriptions apply to a function $f(x)$ as $x \to \infty$. For example,

Definition. $f(x) \to \infty$ as $x \to \infty$ if

$$A; \; \exists X. \; f(x) > A \quad for \; all \; x > X.$$

You should also define, following §2.3, the meaning of

$$f(x) \to l \quad as \quad x \to \infty.$$

(We singled out null sequences as forming the simplest introduction to limits of sequences. Now we need not move so slowly, and can omit special mention of 'null functions' for which $l = 0$.)

The variable x may take arbitrarily large negative values, and we have, for example,

Definition. $f(x) \to l$ as $x \to -\infty$ if, given ϵ, there is X such that

$$|f(x) - l| < \epsilon \quad for \; all \; x < -X.$$

There is a further limit situation which does not occur for sequences s_n, but which presents itself for functions of x. To exemplify it, consider

$$y = \frac{1}{x-2}.$$

As x approaches 2, taking values greater than 2, $y \to \infty$. (Also, as x approaches 2 through values less than 2, $y \to -\infty$.) We can denote the approach of x to a number c through values greater than c by writing $x \to c+$ (and through values less than c by $x \to c-$).

Definition. $f(x) \to \infty$ as $x \to c+$ if, given A, there is δ such that

$$f(x) > A \quad for \; all \; x \; in \quad c < x < c+\delta.$$

3.3. Sketching of curves

It is a good exercise in appreciation of functional dependence to sketch some simple curves. The reader should aim at determining, with hardly any calculation, the *general shape* of the curve. By asking himself the right questions he can do this very quickly. (If a more accurate graph is required, points can be plotted later in the usual way.)

Information under the following headings gives a good start:

(1) the range of x for which y is defined, and any simplifying features such as symmetry about an axis;

(2) values of y when x is large (horizontal asymptotes);

(3) any values of x for which y is large (vertical asymptotes);

(4) any particular points on the curve which can be seen at a glance, e.g. points on the axes.

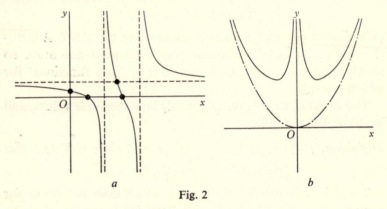

a   b

Fig. 2

Example 1. $y = \dfrac{(x-1)(x-3)}{(x-2)(x-4)}$.

(1) y is defined for all x except 2 and 4.

(2) As $x \to \infty$ or $x \to -\infty$, $y \to 1$. Draw the horizontal asymptote $y = 1$. Also it is clear that, if x is large and positive, $x-1 > x-2$ and $x-3 > x-4$, so the curve approaches the asymptote from above as $x \to \infty$.

(3) $x = 4$ and $x = 2$ are vertical asymptotes. From the signs of the factors in the numerator and denominator,

$$\text{as} \quad x \to 4+, \quad y \to +\infty \quad \text{and as} \quad x \to 4-, \quad y \to -\infty;$$
$$\text{as} \quad x \to 2+, \quad y \to +\infty \quad \text{and as} \quad x \to 2-, \quad y \to -\infty.$$

(4) $y = 0$ gives $x = 1$ or 3,⎫
$x = 0$ gives $y = \frac{3}{8}$.   ⎬ Mark these points.
                                        ⎭

It is often worth while to find where the curve cuts its horizontal asymptote. Here $y = 1$ gives $x = \frac{5}{2}$ (and, what is more important, no other point).

We have by now a good idea of the shape of the curve (fig. 2a).

Example 2.
$$y = x^2 + \frac{1}{x^2}.$$

(1) Since y is unchanged when $-x$ is put in place of x, the curve is symmetrical about Oy. y is defined and is positive for all x except 0.

(2) As $x \to +\infty$ or $x \to -\infty$, $y \to \infty$. Moreover, when x is large, y is a little greater than x^2, so we sketch the parabola $y = x^2$ as a guide for large x.

(3) As $x \to 0+$ or $x \to 0-$, $y \to \infty$.

(4) The curve does not cut either axis. When $x = \pm 1$, $y = 2$, (see fig. 2b).

Exercises 3 (*a*)

Sketch the general shapes of the curves given by the following equations.

1. $y = x^{10}$.   **2.** $y = x^{-10}$.   **3.** $y = x^{10} + x^{-10}$.

4. $y = \dfrac{x^2}{x+1}$.   **5.** $y = \dfrac{1}{x} - \dfrac{3}{x^2}$.

6. $y = \dfrac{x^2+1}{(x-2)(x-4)}$.   **7.** $y = \dfrac{(2x-5)(x-3)}{(x-2)(x-4)}$.

8. $y^2 = \dfrac{x^2}{x^2+1}$.   **9.** $y^2 = \dfrac{x^2}{x+1}$.

3.4. Continuous functions

The reader will have acquired from examples the impression that the common functions can reasonably be called *continuous*, though some of them may present discontinuities for particular values of x. For instance, he would regard the function whose graph is sketched in example 1 of §3.3 as being continuous except at $x = 2$ and $x = 4$. He would think of a function as continuous so long as its graph can be drawn without taking the pencil off the paper.

We must now refine these rough ideas into analytical concepts. The reader, after reflection, will (we hope) agree that for a function f to be continuous at a value $x = c$ he requires that (1) $f(c)$ is defined, (2) as x approaches c, the value of $f(x)$ approaches $f(c)$. Thus the assertion of continuity is nothing more or less than a statement about limits, namely that

$$\lim_{x \to c+} f(x) = \lim_{x \to c-} f(x) = f(c).$$

We need not keep the $c+$ and $c-$ separate and can now set up two basic definitions, the former dealing with the limit of a function and the latter with continuity.

Definition. $f(x) \to l$ *as* $x \to c$ *if, given* ϵ*, there exists* δ *such that*

$$|f(x)-l| < \epsilon \quad \text{for all } x \text{ in} \quad 0 < |x-c| < \delta.$$

The δ depends on the ϵ, and, in general, the smaller the value of ϵ, the smaller must be the value of δ. The dependence may be emphasised if we wish, by writing $\delta(\epsilon)$.

Note that the value $x = c$ is excluded from the set of x for which the ϵ-inequality is required to hold in the above definition. The statement concerns behaviour as x gets near to c, not behaviour at c.

Definition. *The function* f *is* continuous at c *if* $f(x) \to f(c)$ *as* $x \to c$.

Combining the last two definitions we see that the following is an equivalent definition of continuity.

f *is continuous at* c *if, given* ϵ*, there is* δ *such that*

$$|f(x)-f(c)| < \epsilon \quad \text{for} \quad |x-c| < \delta.$$

The values of x such that $|x-c| < \delta$, in other words, the open interval $(c-\delta, c+\delta)$ may be aptly called a *neighbourhood* of c. More generally we can apply this word to an open interval $c-\delta_1 < x < c+\delta_2$, where δ_1, δ_2 may be different. It is worth while to state this as a definition.

Definition. *An open interval is called a* neighbourhood *of any one of its points.*

The definition of continuity of f at c then states that, given any neighbourhood N_1 of $f(c)$, there is a neighbourhood N_2 of c such that, if x is in N_2, then $f(x)$ is in N_1.

We have defined continuity of f at c, that is to say, continuity at a point. We go on to define continuity in an interval. Suppose first that the interval is closed.

Definition. f *is said to be continuous in the closed interval* (a, b) *if*

(1) *for each* c *in* $a < c < b$, f *is continuous at* c;

(2) $\lim\limits_{x \to a+} f(x) = f(a), \quad \lim\limits_{x \to b-} f(x) = f(b).$

The purpose of the special treatment of the end-points a and b is clear; we wish to avoid any mention of values of x outside (a, b). If we are defining continuity in an open interval say $a < x < b$ or the infinite interval $x > a$, there are no end-points and no condition (2).

Definition. f *is said to be continuous in an open interval if it is continuous at each point of the interval.*

3.5. Examples of continuous and discontinuous functions

We shall amplify and make more precise the remark made at the beginning of §3.4 that the common functions of x are generally continuous.

Bounds of f. Suppose that x is allowed to take any value in a set X. In practice X is usually an interval, which may be closed or open. The values of $f(x)$ for x in X form a set of numbers, Y (sometimes denoted by $f(X)$). If the set Y is bounded (§1.7) we say that the function f is bounded in X. Also the sup and inf of Y are called the sup and inf of the function f for x in X.

Continuous functions. In the following theorems, the condition of continuity may refer to a point or to an interval (closed or open) so long as it is given the same interpretation in the hypothesis and the conclusion.

The sum of two continuous functions is continuous.

The product of two continuous functions is continuous.

The quotient of two continuous functions is continuous for any value of x for which the denominator is not zero.

The reader should satisfy himself that these constantly used facts can be proved by arguments following those set out for the fundamental theorems on sequences given in §2.5.

We have further

If n is a positive integer, the function x^n is continuous for all values of x; x^{-n} is continuous except for $x = 0$.

This can be proved directly (by proving that $x^n - c^n$ is small if $x - c$ is small) or alternatively by induction, applying the theorem about the product of continuous functions to x^{n-1} and x.

We can now build up sums of multiples of powers of x to give

Any polynomial is continuous for all values of x.

A quotient of two polynomials is continuous for all values of x for which the denominator is not zero.

A more general and much-used result is the following, which asserts the continuity of the function obtained by compounding two continuous functions.

Theorem 3.5. *Suppose that*
  (1) $g(x)$ *is continuous for* $x = \xi$, *and* $g(\xi) = \eta$;
  (2) $f(y)$ *is continuous for* $y = \eta$.
Then $f\{g(x)\}$ is continuous for $x = \xi$.
  Proof. Write
$$g(\xi + h) = \eta + k.$$

Given ζ, we can find δ such that

$$|k| = |g(\xi + h) - g(\xi)| < \zeta \quad \text{if} \quad |h| < \delta.$$

Given ϵ, we can find ζ such that

$$|f(\eta + k) - f(\eta)| < \epsilon \quad \text{if} \quad |k| < \zeta.$$

These statements combine to give

$$|f\{g(\xi + h)\} - f\{g(\xi)\}| < \epsilon \quad \text{if} \quad |h| < \delta. \ |$$

Some discontinuous functions. For continuity at c it is necessary and sufficient that $f(c+) = f(c) = f(c-)$, where $f(c+)$ is written for the limit of $f(x)$ as $x \to c+$. Examples may be constructed in which one or more of these necessary conditions fail, giving a discontinuity at c. We give two illustrations.

(1) Let $f(x) = [x] = $ the greatest integer less than or equal to x.

This function is continuous if x is not an integer. If x is a positive integer n, $f(n+) = f(n) = n$, $f(n-) = n-1$.

(2) $f(x) = \sin(1/x)$. This function is not defined for $x = 0$. If the definition is completed by assigning a value to $f(0)$, the function would be discontinuous for $x = 0$, whatever the value of $f(0)$. For $f(x)$ does not tend to a limit as $x \to 0$, since it takes all values between -1 and 1 (inclusive) for values of x as near to 0 as we wish, e.g. if $(1/x) = (2n + \frac{1}{2})\pi$, $\sin(1/x) = 1$ and by taking n large enough this value of x is arbitrarily close to 0.

Exercises 3 (*b*)

Notes on these exercises are given on pp. 174–5.

1. State any values of x for which the following functions are discontinuous

$$\sqrt{\{(x-a)/(b-x)\}}, \quad 1/\sqrt{(x^3+1)},$$

$$\tan x, \quad \sec x, \quad 1/(1+\tan x).$$

2. Sketch the graph of the 'saw-tooth function'

$$x-[x]-\tfrac{1}{2}.$$

3. Sketch the graphs of the functions

$$[x^2], \quad [\sqrt{x}], \quad \sqrt{[x]},$$

pointing out for what values of x there are discontinuities.

4. Prove that the function defined by

$$f(x) = x \sin (1/x) \quad (x \neq 0),$$

$$f(0) = 0$$

is continuous for all values of x. Sketch its graph.

5. Prove that, if

$$f(x) = \frac{x^2-6x+5}{x^2-9x+18},$$

and $k \neq 1$, there are two values of x for which $f(x) = k$. Illustrate by a graph.

6. Prove that

$$\frac{x^2-2x+1}{x^2+2x+2}$$

is bounded for all values of x, and find its supremum and infimum.

7. Answer the same question for

$$\frac{4x^2+3}{x^4+1}.$$

8. Investigate the limit as $x \to 1$ of

$$\frac{x^2-3x+2}{x^3-3x^2+2}.$$

9. Investigate the limits as $x \to 0$ of

(i) $\dfrac{\sqrt{(1+x)}-\sqrt{(1-x)}}{x}$,   (ii) $\dfrac{a_0+a_1x+...+a_mx^m}{b_0+b_1x+...+b_nx^n}$,

(iii) $\dfrac{a_0x^p+a_1x^{p+1}+...+a_mx^{p+m}}{b_0x^q+b_1x^{q+1}+...+b_nx^{q+n}}$ $(a_0 \neq 0, b_0 \neq 0)$.

10. Find numbers p, q, r such that

$$p+\frac{q}{x}+\frac{r}{x^2}$$

gives the closest possible approximation for large values of x to the given functions

$$\text{(i)} \quad \frac{x}{x^2+x+1}, \quad \text{(ii)} \quad \frac{x}{\sqrt{(x^2+4)}}.$$

3.6. The intermediate-value property

Theorem 3.6. *Suppose that f is continuous in the closed interval (a, b) and that $f(a) \neq f(b)$. Then f takes every value which lies between $f(a)$ and $f(b)$.*

Remarks. Suppose that $f(a) < f(b)$ and that η is a number such that $f(a) < \eta < f(b)$. The theorem asserts that the curve $y = f(x)$ cuts the line $y = \eta$, i.e. that there is a number ξ between a and b for which $f(\xi) = \eta$. (There may be more than one such ξ.) Taking the intuitive view that a continuous function is one whose graph can be drawn without lifting the pencil off the paper, we can have no doubt that the theorem is true.

We must, however, provide an analytical proof, and a little reflection shows a likely approach to one. Let x take values increasing from a to b. While x is near enough to a, $f(x)$ will still be less than η. When x is near enough to b, $f(x)$ is greater than η. If we take the supremum ξ of the numbers x for which $f(x) < \eta$, we may hope to prove that $f(\xi) = \eta$.

Alternatively we could have expressed this groping for a proof in the language of a Dedekind section. However, we have at our disposal theorem 1.8, asserting the existence of the supremum (which was proved by a Dedekind section), and it is more economical to use it. The formal details of the proof now follow.

Proof of theorem 3.6. Let S be the set of numbers x in $a \leqslant x \leqslant b$ for which $f(x) < \eta$. All the numbers of S are less than b. By theorem 1.8, they have a supremum ξ, where $\xi \leqslant b$.

The theorem will be proved when we have shown that

$$\text{(i)} \quad f(\xi) \leqslant \eta,$$

and $$\text{(ii)} \quad f(\xi) \geqslant \eta.$$

We prove (i). By the definition of supremum, there are values of x in S arbitrarily near to ξ. For these $x, f(x) < \eta$. By continuity of $f(x)$ at ξ, $f(\xi)$ is the limit of these values $f(x)$ and therefore

$$f(\xi) \leqslant \eta.$$

We now prove (ii). Observe first that ξ cannot be b, for, if so, the last inequality would give $f(b) \leqslant \eta$, which is false. So $\xi < b$.

By continuity, $f(\xi)$ is the limit of $f(x)$ as x approaches ξ through values greater than ξ. None of these values are in S and so, for them, $f(x) \geqslant \eta$.

Therefore $f(\xi) = \lim f(x) \geqslant \eta$. This is (ii). **∎**

3.7. Bounds of a continuous function

We go on to establish other general properties of continuous functions.

Theorem 3.71. *If f is continuous in the closed interval (a, b), it is bounded in (a, b).*

Remarks. It is essential to the truth of the theorem that the interval should be closed. The function $1/x$ is continuous in the interval $0 < x \leqslant 1$, open at the left-hand end. By choosing x near enough to 0, we can make $1/x$ arbitrarily large, so it is not bounded above in $0 < x \leqslant 1$.

We shall establish the existence of an upper bound for f, i.e. a number K such that $f(x) \leqslant K$ for $a \leqslant x \leqslant b$. A corresponding argument shows the existence of a lower bound. The proof has some likeness to the proof of the intermediate-value theorem, in that it uses the supremum of a suitably defined set of values x in (a, b).

Proof. Let S be the set of numbers x_1 in $a \leqslant x_1 \leqslant b$ such that the function-values $f(x)$ are bounded above for $a \leqslant x \leqslant x_1$. Then S is not empty since a belongs to S; and the members of S are less than or equal to b. Therefore the members of S have a supremum ξ. There are three possibilities

$$\text{(i)} \ a < \xi < b, \quad \text{(ii)} \ \xi = a, \quad \text{(iii)} \ \xi = b.$$

We shall prove that (i) and (ii) lead to contradictions. We shall obtain the contradiction in (i) by producing a member of S

which is greater than ξ. Since f is continuous at ξ, there is an interval $(\xi-\delta, \xi+\delta)$ inside (a, b) within which

$$f(x) < f(\xi)+1.$$

(Any positive number in place of 1 would serve just as well—if we take the appropriate δ.) Since ξ is the supremum of the numbers in S, we can find K and an x_1 of S for which

$$f(x) < K \quad \text{for} \quad a \leqslant x \leqslant x_1$$

and
$$x_1 > \xi-\delta.$$
Therefore

$$f(x) < \max\{K, f(\xi)+1\} \quad \text{for} \quad a \leqslant x \leqslant \xi+\tfrac{1}{2}\delta.$$

So the number $\xi+\tfrac{1}{2}\delta$ is in S, which contradicts the definition of ξ as the supremum of numbers in S.

We prove next that the assumption (ii) $\xi = a$ leads to a contradiction.

Since f is continuous (to the right) at a, we can find δ such that

$$f(x) < f(a)+1$$

for $a \leqslant x < a+\delta$.

So $a+\tfrac{1}{2}\delta$ is in S which contradicts the assumption $\xi = a$.

We are therefore left with (iii) $\xi = b$, and we have to deduce from this that f is bounded above for $a \leqslant x \leqslant b$.

Since f is continuous at b, there is a δ such that

$$f(x) < f(b)+1 \quad \text{for} \quad b-\delta < x \leqslant b.$$

Since b is the supremum of numbers in S, there is K and an x_1 of S for which

$$f(x) < K \quad \text{for} \quad a \leqslant x \leqslant x_1$$

where
$$x_1 > b-\delta.$$
Therefore
$$f(x) < \max\{K, f(b)+1\} \quad \text{for} \quad a \leqslant x \leqslant b. \; |$$

Theorem 3.72. *A function continuous in a closed interval attains its bounds. In symbols, if f is continuous for $a \leqslant x \leqslant b$ and $M = \sup f(x)$, then there is a value x_1 in $a \leqslant x_1 \leqslant b$ for which $f(x_1) = M$ (with a similar statement for $m = \inf f(x)$).*

Remarks. Just as for the last theorem, it is essential that the interval should be closed. For instance, the function $x+2$ is continuous; its bounds in the open interval $0 < x < 1$ are $M = 3$ and $m = 2$; there are no values of x in $0 < x < 1$ for which the function $x+2$ takes these values.

We give two proofs. The first is short, but rather tricky. The second is an illustration of a very powerful general technique— the method of bisection—which the student may well master now.

First proof. Suppose that there is no x for which

$$a \leqslant x \leqslant b \quad \text{and} \quad f(x) = M.$$

Then $\qquad M - f(x) > 0 \quad \text{for all } x \text{ in} \quad a \leqslant x \leqslant b.$

From §3.5 the function

$$\frac{1}{M - f(x)}$$

is continuous for $a \leqslant x \leqslant b$, since the denominator does not vanish.

So by the last theorem it is bounded and there is k such that

$$\frac{1}{M - f(x)} < k \quad \text{for} \quad a \leqslant x \leqslant b.$$

This gives $\qquad f(x) < M - \dfrac{1}{k} \quad \text{for} \quad a \leqslant x \leqslant b,$

which contradicts $\sup f(x) = M$. ▮

Second proof (by the bisection method). Bisect the interval (a, b). The key to the proof is that there is at least one of the two halves in which $\sup f(x) = M$. Select this half (or, if $\sup f(x) = M$ in both halves, select (say) the left-hand half).

We thus have an interval, which we letter (a_1, b_1), where either $a_1 = a$ or $b_1 = b$, such that

$$b_1 - a_1 = \tfrac{1}{2}(b - a).$$

Now bisect the interval (a_1, b_1) and repeat the argument. We obtain an interval (a_2, b_2) in which $\sup f(x) = M$, where

$$b_2 - a_2 = \tfrac{1}{4}(b - a).$$

Continue the process indefinitely. It gives an infinite sequence of intervals (a_n, b_n), in every one of which $\sup f(x) = M$, such that

$$a \leqslant a_1 \leqslant a_2 \leqslant \ldots,$$
$$b \geqslant b_1 \geqslant b_2 \geqslant \ldots,$$
$$b_n - a_n = (b-a)/2^n.$$

The last three lines show that the increasing sequence a_n and the decreasing sequence b_n tend to the same limit, ξ say. (Note that ξ may be a or b.)

It is easy to see that $f(\xi) = M$. A formal proof is:

Suppose, if possible, that $f(\xi) = k < M$. Choose k_1 such that $k < k_1 < M$. By continuity of $f(x)$ at ξ, we can find an interval $(\xi - \delta, \xi + \delta)$ in which $f(x) < k_1$. But, if n is large enough, the interval (a_n, b_n) lies inside the interval $(\xi - \delta, \xi + \delta)$. The inequality $f(x) < k_1$ contradicts the fact that, in (a_n, b_n), $\sup f(x) = M$. ▮

3.8. Uniform continuity

Suppose that, in an interval (a, b),

$$\sup f(x) = M \quad \text{and} \quad \inf f(x) = m.$$

The number $M - m$ merits a descriptive word and we shall call it the *leap* of the function in the interval. ('Oscillation' is sometimes used but it has the drawback of suggesting a wave-like function.)

In this book the theorems of this section will be needed only in chapter 7. Theorem 3.82 will then be found to be vital in setting up the integral of a continuous function.

Theorem 3.81. *Suppose that f is continuous in the closed interval (a, b). Then, given ϵ, the interval can be divided into a finite number of parts in each of which the leap of f is less than ϵ.*

Remark. Suppose that c is the mid-point of (a, b). If both of the intervals (a, c), (c, b) can be divided into a finite number of parts in which the leap of f is less than ϵ, these parts form a subdivision of (a, b) with the same property. So the theorem lends itself to proof by bisection.

Proof. Suppose the theorem false. Bisect the interval (a, b). There is at least one half for which it is false. Choose this half (or the left-hand half, if it is false for both halves) and denote it by (a_1, b_1). Repeat the bisection. We have an increasing sequence a_n and a decreasing sequence b_n with a common limit ξ.

Since f is continuous at ξ, there is an interval $(\xi - \delta, \xi + \delta)$ in which the leap of f is less than ϵ. (If ξ is a or b, the interval is $(a, a + \delta)$ or $(b - \delta, b)$, respectively.)

But, if n is large enough, the interval $(\xi - \delta, \xi + \delta)$ includes (a_n, b_n). From the definition of (a_n, b_n) it cannot be divided into a finite number of parts in which the leap of f is less than ϵ. This is a contradiction. ∎

Theorem 3.82. *Let f be continuous in the closed interval (a, b). Given ϵ, there is δ such that, if x_1 and x_2 are any two points of (a, b) with $|x_1 - x_2| < \delta$, then*
$$|f(x_1) - f(x_2)| < \epsilon.$$

Proof. From the last theorem, we can divide (a, b) into a finite number of subintervals in each of which the leap of $f(x)$ is less than $\frac{1}{2}\epsilon$. Take δ to be the length of the smallest of these subintervals. If $|x_1 - x_2| < \delta$, then x_1 and x_2 lie in the same subinterval or in adjacent ones. In the former case
$$|f(x_1) - f(x_2)| < \tfrac{1}{2}\epsilon.$$

In the latter, if c is the common end-point of the two subintervals,
$$|f(x_1) - f(x_2)|$$
$$\leqslant |f(x_1) - f(c)| + |f(c) - f(x_2)|$$
$$< \tfrac{1}{2}\epsilon + \tfrac{1}{2}\epsilon = \epsilon. \quad \blacksquare$$

This is called the theorem of uniform continuity. The reader may treat this phrase merely as a label until (beyond the scope of this book) he comes across the notion of uniformity in other contexts in analysis.

The point of the theorem can be conveyed by the following remarks. Continuity for a value x ensures that, given ϵ, there is δ such that, if x_1 and x_2 are any points in the interval $(x - \delta, x + \delta)$, then $|f(x_1) - f(x_2)| < \epsilon$. This δ depends on ϵ and also on the particular value x. Theorem 3.82 states that it is possible to choose a δ which will serve for every x in (a, b).

3.9. Inverse functions

You will have met situations in which, uncritically, a function has been constructed as the *inverse* of an already known function. For example, the equation $x = \sin y$ is used to define y as a function arc sin x of x. This illustration shows that the procedure requires care, because, for a given x, there are infinitely many values of y, and if we are to keep to single-valued functions we must impose some restriction on the permitted values of y.

We shall now give an *existence theorem* which assures us that, if certain simple conditions are fulfilled, we can obtain a new function inverse to a known function.

Definition. f is increasing *for $a \leqslant x \leqslant b$ if $f(x_1) \leqslant f(x_2)$ for all x_1, x_2 such that $a \leqslant x_1 < x_2 \leqslant b$. If $f(x_1) < f(x_2)$, we say that f is* strictly increasing.

Theorem 3.9. *Let f be continuous and strictly increasing for $a \leqslant x \leqslant b$. Let $f(a) = c$, $f(b) = d$. Then there is a function g, continuous and strictly increasing for $c \leqslant y \leqslant d$, such that $f\{g(y)\} = y$ (so that $g(y)$ is the function inverse to $f(x)$).*

Remarks. The situation is easily appreciated graphically. Moreover, we can see that the hypotheses are natural. If f were not strictly increasing (or strictly decreasing) there could be values of y corresponding to more than one value of x, and then there would not be a single-valued inverse. If f were discontinuous, we might have, for some ξ, $f(\xi+) = \mu > \lambda = f(\xi-)$ and an interval (λ, μ) of y in which there is no inverse function.

Proof. Let k be any number such that $c < k < d$.

By theorem 3.6, there is a value h such that

$$f(h) = k,$$

and, since f strictly increases, there is only one such h corresponding to a given k.

The inverse function g is then defined by $h = g(k)$.

It is easy to see that g is strictly increasing. For a formal proof of this, let $y_1 < y_2$ and $y_1 = f(x_1)$, $y_2 = f(x_2)$. From the last paragraph, x_1 and x_2 are uniquely defined. If $x_2 \leqslant x_1$, then, since f is increasing, $f(x_2) \leqslant f(x_1)$, i.e. $y_2 \leqslant y_1$, which contradicts the assumption $y_1 < y_2$. So $x_1 < x_2$ and g is strictly increasing.

It remains to prove that g is continuous.

With $\epsilon > 0$, let

$$f(h-\epsilon) = k_1 \quad \text{and} \quad f(h+\epsilon) = k_2.$$

Then, since f is increasing,

$$k_1 < k < k_2$$

and $h-\epsilon < g(y) < h+\epsilon$ if $k_1 < y < k_2.$

Since ϵ is arbitrary, g is continuous at $y = k$.

Here k is any number in the open interval (c, d). A similar argument establishes one-sided continuity at the end-points c and d.

Exercises 3 (c)

Notes on these exercises are given on p. 175.

1. Discuss the continuity of the following functions:
    (i) $f(x) = 1/q$ when x is a rational number p/q in its lowest terms, $f(x) = 0$ when x is irrational;
    (ii) $f(x) = x \log \sin^2 x \ (x \neq 0), f(0) = 0$;

    (iii) $f(x) = \dfrac{1}{x-a} \operatorname{cosec} \dfrac{1}{x-a}.$

2. Construct a function which takes every value y in $0 \leqslant y \leqslant 1$ once and once only for values of x in $0 \leqslant x \leqslant 1$ and which is discontinuous for some values of x in $0 \leqslant x \leqslant 1$.

3. f is bounded for $a \leqslant x \leqslant b$, and $M(x)$ is the supremum of the values of f in the closed interval (a, x). Prove that, if f is continuous for a value x_0 in $a < x_0 < b$, and if $f(x_0) < M(x_0)$, then there is an interval containing x_0 in which $M(x)$ is constant.

4. Can a function be continuous for one value of x and discontinuous for all other values?

5. In theorem 3.82, take $f(x) = x^{1/3}, a = -1, b = 1, \epsilon = \frac{1}{10}$ and give a value of δ.

6. Which of the following sets of data (i)–(iii) are sufficient to determine the value $f(0)$?
    (i) f is continuous at $x = 0$ and in any neighbourhood of $x = 0$ it takes both positive and negative values;

(ii) given ϵ, there is δ such that $|f(x)| < \epsilon$ for $0 < |x| < \delta$;

(iii) as $h \to 0$, $f(h) \to a$ and

$$\frac{f(h)+f(-h)-2f(0)}{h} \to l.$$

7. f is bounded for $a \leqslant x \leqslant b$ and, for every pair of values x_1, x_2 with $a \leqslant x_1 \leqslant x_2 \leqslant b$,

$$f\{\tfrac{1}{2}(x_1+x_2)\} \leqslant \tfrac{1}{2}\{f(x_1)+f(x_2)\}.$$

Prove that f is continuous for $a < x < b$.

8. f is defined for all x and

$$|f(x)-f(x')| \leqslant \alpha|x-x'|$$

for all x and x', where the constant α is less than 1. The sequence $x_1, x_2, \ldots$ is defined by a given x_0 and

$$x_n = f(x_{n-1}) \quad (n = 1, 2, \ldots).$$

Prove that $x_n \to \xi$ and $n \to \infty$, where

$$\xi = f(\xi).$$

Prove also that, if $f(0) \geqslant 0$, then

$$\frac{f(0)}{1+\alpha} \leqslant \xi \leqslant \frac{f(0)}{1-\alpha}.$$

4

THE DIFFERENTIAL CALCULUS

4.1. The derivative

If, for a given value of x,

$$\frac{f(x+h)-f(x)}{h}$$

tends to a finite limit as h tends to 0, this limit, denoted by $f'(x)$, is called the *derivative* or *differential coefficient* of f for the value of x. The function f is said to be *differentiable* for that value of x.

To state the geometrical counterpart of this definition, suppose that P is a given point on the curve $y = f(x)$ and Q is a variable point on the curve. If, as Q approaches P along the curve, the gradient of the line PQ tends to a limit, the curve has a tangent at P whose gradient is $f'(x)$.

The definition calls for a number of remarks.

(1) An expressive notation is to write h, the change in x, as δx (δx is a single symbol; it is not δ multiplied by x). The corresponding change in y, namely, $f(x+\delta x)-f(x)$ is called δy. The derivative is then the limit, as δx tends to 0, of the ratio

$$\frac{\delta y}{\delta x}.$$

The derivative may be written

$$\frac{dy}{dx},$$

where d/dx is, for the present, to be understood as a symbol specifying an operation to be performed on whatever follows the d in the upper line. (The dy and dx are not the numerator and denominator of a fraction.)

(2) A function, defined for a set of values of x, may be differentiable for all those values or for none or for some and not others.

Illustrations. (i) $f(x) = x^2$.

$$f'(x) = 2x \quad \text{for all values of } x.$$

(ii)
$$f(x) = x \quad (x \text{ rational}),$$
$$= 0 \quad (x \text{ irrational}),$$

$f(x)$ is not differentiable for any value of x.

(iii)
$$f(x) = |x|.$$

$f'(x) = 1$ if $x > 0$ and $f'(x) = -1$ if $x < 0$. f is not differentiable for $x = 0$.

(3) A necessary (but not sufficient) condition for f to be differentiable for a given value of x is that f is continuous for that value.

For $\delta y/\delta x$ can tend to a finite limit as $\delta x \to 0$ only if $\delta y \to 0$, i.e. if f is continuous.

The function $|x|$ is continuous but not differentiable at $x = 0$.

(4) The following notes are, in part, suggested by the examples in (2).

In the definition of the derivative we could, if we wished, consider separately $h \to 0+$ and $h \to 0-$, defining what might be called the right-hand and left-hand derivatives. If they are equal there is a derivative in the ordinary sense $f'(x)$. The function $|x|$ of (iii) above has at $x = 0$ right-hand derivative 1 and left-hand derivative -1.

Consider next another example of a power of x. If $f(x) = x^{\frac{1}{3}}$, then, for $x = 0$,
$$\frac{f(h) - f(0)}{h} = h^{-\frac{2}{3}}$$

and this tends to $+\infty$ as $h \to 0$. Geometrically, the curve $y^3 = x$ has a vertical tangent at $(0, 0)$. In defining the derivative we specified that the limit should be finite, and in accordance with this definition we say that $x^{\frac{1}{3}}$ is not differentiable for $x = 0$.

The decision whether to admit or exclude infinite derivatives is based on convenience. The exceptional cases that have to be stated in theorems if infinite derivatives are allowed sway the balance in favour of exclusion. For instance, if, for a certain value of $x, f'(x)$ were $+\infty$ and $g'(x)$ were $-\infty$, it can be shown by simple examples that $f+g$ might have a derivative with any value or need not be differentiable; so the rule to be given in §4.2 (1) for differentiating the sum of two functions would lose the simplicity that it has for finite derivatives.

Exercises 4 (a)

Notes on these exercises are given on p. 175.

1. Find the equation of the tangent to the curve $y = x^2 - 4$ at each of its points of intersection with (i) the x-axis, (ii) the y-axis.

2. Find the equation of the normal to the curve $2y = (x+2)^2$ at each of the points on it where $y = 2$.

3. Find the equation of the tangent at the origin to the curve

$$y = x^2(x-1)^2 + 3x.$$

Prove that the tangent touches the curve again at a second point.

4. Prove that the equation of the tangent to the parabola $y^2 = 4ax$ at the point $(at^2, 2at)$ is

$$ty = x + at^2.$$

5. Give an example of a function which is continuous for all values of x and is differentiable for all values of x except 1 and -1.

6. State for what values of x the following functions fail to be (a) continuous, (b) differentiable.

(i) $[x]$, (defined on p. 54).

(ii) $f(x) = \begin{cases} x-1 & (x \leqslant 1), \\ x(x-1) & (x > 1). \end{cases}$

7. Find the angles of intersection of the pair of curves

(i) $2y = x^2, \quad y = x^2(2-x)$.

Also of the pair

(ii) $2y = x^2, \quad 2x = y^2$.

4.2. Differentiation of sum, product, etc.

We assume that the functions f and g have derivatives for the values of x considered. The proof of (1) is left to the reader.

(1) If $s(x) = f(x) + g(x)$, then $s'(x) = f'(x) + g'(x)$.
If $t(x) = kf(x)$, then $t'(x) = kf'(x)$.

(2) If $\phi(x) = f(x)\,g(x)$, then $\phi'(x) = f(x)\,g'(x) + f'(x)\,g(x)$.
Proof of (2). We have

$$\frac{\phi(x+h) - \phi(x)}{h} = \frac{f(x+h)\,g(x+h) - f(x)\,g(x)}{h}$$

$$= f(x+h)\frac{g(x+h) - g(x)}{h} + g(x)\frac{f(x+h) - f(x)}{h}.$$

Consider the first term in this last sum. Since f is differentiable at x it is continuous at x and so $f(x+h) \to f(x)$ as $h \to 0$. The other factor $\{g(x+h)-g(x)\}/h$ has limit $g'(x)$. Since the limit of a product is equal to the product of the limits,

$$f(x+h)\frac{g(x+h)-g(x)}{h} \to f(x)\,g'(x) \quad \text{as} \quad h \to 0.$$

Similarly
$$g(x)\frac{f(x+h)-g(x)}{h} \to g(x)\,f'(x) \quad \text{as} \quad h \to 0.$$

Finally, we appeal to the theorem that the limit of the sum of these two functions of h is the sum of the limits. ▮

This formula for differentiating a product is typical of those which are used over and over again in the differential calculus. The proof has been set out fully to stress its dependence on the repeated application of the theorems about limits.

(3) *If* $\phi(x) = 1/g(x)$ *and* $g(x) \neq 0$, *then*

$$\phi'(x) = -\frac{g'(x)}{\{g(x)\}^2}.$$

This follows from

$$\frac{\phi(x+h)-\phi(x)}{h} = \frac{g(x)-g(x+h)}{hg(x)\,g(x+h)}.$$

(4) *If* $\phi(x) = f(x)/g(x)$, *then*

$$\phi'(x) = \frac{f'(x)\,g(x)-f(x)\,g'(x)}{\{g(x)\}^2}.$$

Combine (2) and (3).

(5) *Function of a function.*

Let $u = g(x)$ and $y = f(u)$, so that

$$y = f\{g(x)\} = \phi(x), \text{ say.}$$

Then $\qquad\qquad \phi'(x) = f'\{g(x)\}\,g'(x),$

or, in other notation, $\qquad \dfrac{dy}{dx} = \dfrac{dy}{du}\dfrac{du}{dx}.$

Proof. Let δx be a change in x, δu the corresponding change in u by the functional relation $u = g(x)$. Let δy be the change in y from $y = f(u)$. Then

$$\delta u = \{g'(x)+\epsilon\}\,\delta x,$$

where the number ϵ (which depends on x and δx) tends to 0 as δx tends to 0. Similarly

$$\delta y = \{f'(u) + \epsilon_1\}\, \delta u,$$

where $\epsilon_1 \to 0$ as $\delta u \to 0$. Here we must observe that, if $g'(x) \neq 0$, δu is different from 0 if δx is small enough; but, if $g'(x) = 0$, δu may possibly take zero values for arbitrarily small δx; if $\delta u = 0$, ϵ_1 fails to be determined by the equation connecting δy and δu, and we then define $\epsilon_1 = 0$.

The expressions for δu and δy give

$$\delta y = \{f'(u) + \epsilon_1\}\{g'(x) + \epsilon\}\, \delta x.$$

Divide by δx and let $\delta x \to 0$. |

(6) *Inverse function.*

Let $y = f(x)$ be continuous and strictly increasing for $a \leqslant x \leqslant b$. If, for a given x in $a < x < b$, $f'(x) \neq 0$, then the inverse function $x = g(y)$ is differentiable for the corresponding value of y and

$$g'(y) = \frac{1}{f'(x)}.$$

Proof. The inverse function exists by theorem 3.9.

If h (not zero) is given, define k by

$$y + k = f(x + h).$$

Then $k \neq 0$ and, if k is given, h is determined uniquely from

$$g(y + k) = x + h.$$

So
$$\frac{g(y + k) - g(y)}{k} = \frac{h}{k} = \frac{h}{f(x + h) - f(x)}.$$

Let $h \to 0$. Then $k \to 0$. |

4.3. Differentiation of elementary functions

If m is rational, x^m has derivative mx^{m-1}, except for (i) $x = 0$ when $m < 1$ and (ii) $x \leqslant 0$ when $m = p/q$ (in its lowest terms) with q an even integer.

Proof. Suppose first that the index is a positive integer n. Then, if $h \neq 0$,

$$\frac{(x + h)^n - x^n}{h} = nx^{n-1} + \tfrac{1}{2}n(n - 1)\, x^{n-2}\, h + \ldots + h^{n-1}.$$

As $h \to 0$, the limit of each of the $n-1$ terms on the right after the first is zero.

If now the index is not a positive integer an argument on these lines is less easy. A binomial expansion would be an infinite series; and we have so far no theorems covering the limit as $h \to 0$ of the sum of *infinitely* many terms each of which tends to 0 as $h \to 0$. We adopt a different approach.

Suppose that the index is a negative integer $-n$. Then

$$\frac{(x+h)^{-n} - x^{-n}}{h} = \frac{x^n - (x+h)^n}{h(x+h)^n x^n}.$$

If $x \neq 0$, the limit as $h \to 0$ of the right-hand side is $-nx^{n-1}/x^{2n}$ or $(-n) x^{-n-1}$, which is what we want.

Suppose finally that m is a rational number p/q, where p and q are integers, $q > 0$, and the cases (i), (ii) are excepted. The easiest proof is by use of §4.2 (6), as follows. Write $y = u^p$, $x = u^q$ so that $y = x^{p/q}$. Then

$$\frac{dy}{dx} = \frac{dy}{du}\frac{du}{dx} = \frac{dy}{du} \Big/ \frac{dx}{du}$$

$$= \frac{pu^{p-1}}{qu^{q-1}} = \frac{p}{q} u^{p-q} = mx^{m-1}. \quad \blacksquare$$

We are now able by §4.2, (1)–(4), to differentiate any polynomial

$$p(x) = a_0 x^n + a_1 x^{n-1} + \ldots + a_n$$

and any rational function of x, namely, a quotient $p(x)/q(x)$ of two polynomials, for any value of x for which $q(x)$ is not zero.

To provide a greater variety of exercises it is convenient to assume a knowledge of the following derivatives

$$\frac{d}{dx} e^x = e^x, \quad \frac{d}{dx} \log x = \frac{1}{x} \quad (x > 0),$$

$$\frac{d}{dx} \sin x = \cos x, \quad \frac{d}{dx} \cos x = -\sin x.$$

A more systematic investigation of the properties of exponential, logarithmic and trigonometric functions (and their derivatives) is best postponed until chapter 6 after further discussion of infinite series.

Exercises 4 (*b*)

Notes on these exercises are given on p. 176.

1. Differentiate

$$\frac{x}{(x^2+a^2)^{1/2}}, \quad \frac{x^2}{x^2+a^2}, \quad \frac{x^3}{(x^2+a^2)^{3/2}}.$$

2. Evaluate the following limits

(i) $\dfrac{x^2-4x+3}{x^2+2x-3}$ as $x \to 1$,

(ii) $\dfrac{x^4-a^4}{x^3-a^3}$ as $x \to a$,

(iii) $\dfrac{\sqrt{(2x+3)}-\sqrt{(x+6)}}{\sqrt{(x+1)}-2}$ as $x \to 3$.

3. Prove by induction that the derivative of x^n is nx^{n-1}, n being a positive integer.

4. If $y_1, y_2, \ldots, y_n$ are functions of x, prove that, if $y = y_1 y_2 \ldots y_n$, then

$$\frac{1}{y}\frac{dy}{dx} = \frac{1}{y_1}\frac{dy_1}{dx} + \ldots + \frac{1}{y_n}\frac{dy_n}{dx},$$

where x has a value for which y is not zero.

5. Prove that, if a polynomial $p(x)$ is divisible by $(x-a)^2$, then $p'(x)$ is divisible by $x-a$.

6. Prove that, if $p'(x)$ is divisible by $(x-a)^{n-1}$ and $p(x)$ is divisible by $x-a$, then $p(x)$ is divisible by $(x-a)^n$.

7. Show how **6** is of use in searching for multiple roots of equations. Illustrate by solving the equation

$$x^6 - 3x^2 + 2 = 0.$$

8. Prove that the equation

$$\frac{1}{x-a} + \frac{1}{x-b} + \frac{1}{x-c} = 0$$

can have a pair of equal roots only if $a = b = c$.

9. Prove that, if $p(x)$ is a polynomial, then between any two roots of $p(x) = 0$ lies a root of $p'(x) = 0$.

(This is Rolle's theorem for polynomials. It will be proved for more general functions in §4.5.)

10. If
$$f(x) = \begin{cases} x^2 \sin(1/x) & x \neq 0, \\ 0 & x = 0, \end{cases}$$

prove that $f'(x)$ exists for all values of x and give the values of $f'(x)$ for $x \neq 0$ and of $f'(0)$. Prove that $f'(x)$ is discontinuous at $x = 0$.

11. The elements of a determinant of order n are functions of x. Prove that its derivative is the sum of the n determinants formed by differentiating the elements of one row only leaving the other rows unaltered.

12. Prove that, if p, q, r are polynomials in x of degree not greater than 3, then

$$\begin{vmatrix} p & q & r \\ p' & q' & r' \\ p'' & q'' & r'' \end{vmatrix}$$

is a polynomial of degree not greater than 3.

4.4. Repeated differentiation

Let $y = f(x)$ be differentiable. The derivative $f'(x)$ may itself be differentiable. Notations for this second derivative of $f(x)$ are d^2y/dx^2 or $f''(x)$. Notice that the existence of the second derivative implies the continuity of the first. The nth derivative may be written $f^{(n)}(x)$. No new principles are involved.

There is a theorem of interest on the nth derivative of a product.

Leibniz's theorem. *If f and g are functions of x having n-th derivatives, then*

$$(fg)^{(n)} = f^{(n)}g + {}_nC_1 f^{(n-1)}g' + \ldots + {}_nC_r f^{(n-r)}g^{(r)} + \ldots + fg^{(n)}.$$

The proof, by induction, is left to the reader. The algebraic lemma

$$_{n+1}C_r = {}_nC_r + {}_nC_{r-1}$$

is required.

Exercises 4 (c)

Notes on these exercises are given on p. 176.

In calculating an nth derivative it is necessary to adopt the method which will give the result in the most compact form.

1. Find the nth derivatives of

$$\frac{x-1}{x^2-4}, \quad \frac{x+1}{(x-2)^2}.$$

2. Show that the value of

$$\left(\frac{d}{dx}\right)^n \frac{x^3}{x^2-1}$$

for $x = 0$ is 0 if n is even and $-n!$ if n is odd and greater than 1.

3. Prove that the nth derivative of $\sin x$ is $\sin(x + \tfrac{1}{2}n\pi)$ and investigate corresponding results for $\cos x$, $\sin kx$ and $e^{ax}\sin bx$.

4. Find the nth derivative of $\sin 3x \sin 5x$. (Do *not* use Leibniz's theorem!)

5. Find the nth derivative of $a/(a^2 - x^2)$.

6. Find the nth derivative of $a/(a^2 + x^2)$.

4.5. The sign of $f'(x)$

Definition. *The function f is said to be* strictly increasing *at c if there is a neighbourhood of c in which*

$$f(x) < f(c) \quad \text{for} \quad x < c$$

and $$f(x) > f(c) \quad \text{for} \quad x > c.$$

There is a corresponding definition of 'strictly decreasing at c'.

Theorem 4.51. *If $f'(c) > 0$, then f is strictly increasing at c.*
  Proof. Since

$$\frac{f(c+h) - f(c)}{h}$$

tends to a limit greater than 0, its value is greater than 0 for all sufficiently small h. That is to say, the numerator and denominator have the same sign. **|**

Exercises. (1) Prove that, if f is strictly increasing at c and $f'(c)$ exists, then $f'(c) \geqslant 0$.
  (2) Give examples of functions satisfying the following conditions:
    (i) f is neither strictly increasing nor strictly decreasing for any value of x in (a, b).
    (ii) f is strictly increasing at c, but it is not true that $f'(c) > 0$.

The following important result is known as Rolle's theorem. Rolle gave it for the particular case of a polynomial (see exercise 7(b), 11).

Theorem 4.52. (*Rolle*). *If*
  (1) *f is continuous in the closed interval $a \leqslant x \leqslant b$,*
  (2) *f' exists in the open interval $a < x < b$,*
  (3) *$f(a) = f(b)$,*
then there is a value c, with $a < c < b$, for which

$$f'(c) = 0.$$

Proof. Let $M = \sup f(x)$ and $m = \inf f(x)$ for $a \leqslant x \leqslant b$. Let $f(a) = k$.

If $M = m = k$, then $f'(c) = 0$ for every c between a and b.

Suppose that either $M > k$ or $m < k$, say the former. By theorem 3.6, there is a value c, with $a < c < b$, for which $f(c) = M$.

By (2) $f'(c)$ exists. We shall prove that $f'(c) = 0$.

If $f'(c) > 0$, by theorem 4.51, f is strictly increasing at c and there are values of x to the right of c for which $f(x) > f(c)$. This contradicts the fact that $M = f(c)$ is the supremum of f.

Similarly, if $f'(c) < 0$, there would be a value of x to the left of c at which $f(x) > M$.

Therefore $f'(c)$ must be equal to 0. ▍

Geometrically, Rolle's theorem states that there is for some value of x between a and b a tangent to the curve $y = f(x)$ which is parallel to the chord joining the points where $x = a$ and $x = b$. In the theorem the chord is horizontal. The next theorem is the extension wherein the chord may have any gradient.

Exercises 4 (d)

Notes on these exercises are given on p. 176.

1. Find a value for c in Rolle's theorem when $f(x)$ is
   (i) $(x-a)^m (x-b)^n$ (m and n positive integers);
   (ii) $\sin (1/x)$ in $(1/n\pi, 1/m\pi)$.

2. If $p(x)$ is a polynomial, prove that there is a root of
$$p'(x) + kp(x) = 0$$
between any two real roots of $p(x) = 0$.

3. If a and b are successive roots of $p(x) = 0$, then the number of roots between a and b of
$$p'(x) + kp(x) = 0$$
(each counted according to its multiplicity) is odd.

4. Prove that the equation
$$\left(\frac{d}{dx}\right)^n (x^2 - 1)^n = 0$$
has n real roots, all different and lying between -1 and 1.

5. What information does a knowledge of the roots of $p'(x) = 0$ give about the roots of $p(x) = 0$?

Prove that the equation

$$1-x+\frac{x^2}{2}-\frac{x^3}{3}+\dots+(-1)^n\frac{x^n}{n} = 0$$

has one real root if n is odd and no real root if n is even.

6. Prove that the equation

$$x^n+kx+l = 0$$

has at most two real roots when n is even and at most three when n is odd.

4.6. The mean value theorem

Theorem 4.61. (*The mean value theorem*). *Let f satisfy the conditions* (1) *and* (2) *of Rolle's theorem. Then there is a value c, with $a < c < b$, for which*

$$f(b)-f(a) = (b-a)f'(c).$$

Proof. Write $\qquad \phi(x) = f(x)-kx$

and choose the constant k to make

$$\phi(b) = \phi(a).$$

This gives $\qquad f(b)-f(a) = k(b-a).$

Rolle's theorem gives c such that $\phi'(c) = 0$, i.e. $f'(c) = k$. ▌

It is often convenient to write the result, with $b = a+h$,

$$f(a+h) = f(a)+hf'(a+\theta h),$$

where θ is some number such that $0 < \theta < 1$.

The mean value theorem has important consequences under further assumptions about $f'(x)$.

Corollary 1. *If $f'(x) = 0$ for all x in $a < x < b$, then $f(x)$ is constant for $a \leqslant x \leqslant b$.*

Proof. If x_1 and x_2 are any two points with

$$a \leqslant x_1 < x_2 \leqslant b,$$

then $\qquad f(x_2)-f(x_1) = (x_2-x_1)f'(x_3),$

where $\qquad x_1 < x_3 < x_2,$

and this is 0. ▌

Corollary 2. *If $f'(x) > 0$ for $a < x < b$, then $f(x)$ is a strictly increasing function in the interval $a \leqslant x \leqslant b$.*

Proof. We have to prove that, if x_1 and x_2 are any points with $a \leqslant x_1 < x_2 \leqslant b$, then $f(x_1) < f(x_2)$.

$$f(x_2) - f(x_1) = (x_2 - x_1) f'(x_3) \quad (x_1 < x_3 < x_2)$$
$$> 0. \ |$$

It is often useful to be able to apply corollary 2 under slightly more general hypotheses. We can allow one value c at which the only requirement is the continuity of f; even the existence of $f'(c)$ is not assumed.

To prove that $f(x)$ is still strictly increasing in (a, b), apply corollary 2 to the two intervals (a, c) and (c, b). If x_1 is in (a, c) and x_2 in (c, b), then $f(x_1) < f(c) < f(x_2)$. $|$

The extended statement holds for any finite number of exceptional points c.

Corollary 2 should be compared carefully with theorem 4.51. It assumes more and proves more. If, as in theorem 4.51, we know only that f is increasing at a point, there need be no interval throughout which f is an increasing function (see exercise 4(f), 13).

The following extension of the mean value theorem to two functions will be useful later.

Theorem 4.62. (*Cauchy's mean value theorem*). *Suppose that both the functions f and g are continuous in the closed interval and differentiable in the open interval (a, b). Suppose that $g'(x)$ is different from 0 for all x in $a < x < b$. Then, for some c with $a < c < b$,*

$$\frac{f(b) - f(a)}{g(b) - g(a)} = \frac{f'(c)}{g'(c)}.$$

Proof. (Observe first that it is not sufficient to apply the mean value theorem to f and g separately, because we get two different c's.)

Define
$$\phi(x) = f(x) - kg(x)$$

and choose the constant k to make $\phi(b) = \phi(a)$, i.e.

$$f(b) - f(a) = k\{g(b) - g(a)\}.$$

By Rolle's theorem, there is c such that

$$\phi'(c) = f'(c) - kg'(c) = 0.$$

Next, $g(b) - g(a) \neq 0$, for, if it were, Rolle's theorem would give a number c_1 for which $g'(c_1) = 0$, which contradicts the hypothesis.

Equate the two values found for k. |

Exercises 4 (e)

Notes on these exercises are given on pp. 176–7.

1. In each of (i)–(iii) find, if possible, a number c satisfying the mean value theorem. In any example in which no c can be found, which of the conditions of the theorem is not satisfied?

$$\text{(i)} \quad f(x) = x(x-2)(x-4) \quad (a = 1, b = 3);$$
$$\text{(ii)} \quad f(x) = 1/x^2 \qquad\qquad (a = -1, b = 3);$$
$$\text{(iii)} \quad f(x) = x^{1/3} \qquad\qquad (a = -1, b = 1).$$

2. Taking $f(x)$ to be (i) x^2, (ii) x^n, prove that, as $h \to 0$, the number θ of the mean value theorem tends to a limit.

3. If $f'(x) \to l$ as $x \to \infty$, prove that $f(x)/x \to l$.

4. Discuss the following argument.

Let $f'(x)$ exist for $a < x < b$. Let $a < c < b$. Then, if $a < c+h < b$, the mean value theorem gives

$$\frac{f(c+h) - f(c)}{h} = f'(c+\theta h).$$

Let $h \to 0$. The left-hand side tends to $f'(c)$. So the limit of $f'(c+\theta h)$, as $h \to 0$, exists and is equal to $f'(c)$. That is to say, $f'(x)$ is continuous at $x = c$.

5. Take in theorem 4.62

$$f(t) = t^2, \quad g(t) = 4t^3 - 3t^4 \quad (a \leqslant t \leqslant b)$$

and find whether a c exists when (i) $a = 0, b = 1$; (ii) $a = -1, b = 2$; (iii) $a = -1, b = \frac{2}{3}$. Illustrate geometrically on the curve $x = t^2$, $y = 4t^3 - 3t^4$.

4.7. Maxima and minima

Definition. *f is said to have a* maximum *at c if there is a neighbourhood of c in which $f(x) < f(c)$ except for $x = c$.*

We define a *minimum* by substituting $>$ for $<$.

A phrase covering either a maximum or a minimum is a *turning value*.

(The use of the words maximum and minimum applied to a continuous function will not be confused with their use in §1.7 for a finite set of numbers.)

Theorem 4.71. *If $f'(c)$ exists, a necessary condition that f has a turning value at c is that $f'(c) = 0$.*

Proof. From theorem 4.51, if $f'(c) > 0$, f is strictly increasing at c. If $f'(c) < 0$, f is strictly decreasing at c. Either of these contradicts c being a turning value. |

Notes. (1) A function may have a turning value for a value of x at which there is no derivative, e.g. $|x|$ has a minimum for $x = 0$.

(2) The condition of the theorem is not sufficient, e.g. if $f(x) = x^3$, $f'(0) = 0$ and f is strictly increasing at $x = 0$.

The following criterion serves to distinguish a maximum from a minimum.

Theorem 4.72. *If there is a neighbourhood of c in which $f'(x) > 0$ for $x < c$ and $f'(x) < 0$ for $x > c$, then f has a maximum for $x = c$.*

Proof. Theorem 4.61, corollary 2, shows that f decreases as x increases from c and also as x decreases from c. |

Alternatively the second derivative may be used to investigate whether a value of c for which $f'(c) = 0$ is a maximum or a minimum of f or neither.

Theorem 4.73. *Let $f'(c) = 0$. If $f''(c) < 0$, then $x = c$ gives a maximum of $f(x)$. If $f''(c) > 0$, $x = c$ gives a minimum.*

Proof. By theorem 4.51 f' is strictly decreasing at c. Therefore there is a neighbourhood of c in which $f'(x) > 0$ for $x < c$ and $f'(x) < 0$ for $x > c$. Apply theorem 4.72. |

Exercises. Investigate maxima and minima of the functions

(i) $|x|$,

(ii) $x/(1+x^2)$,

(iii) $(ax+b)/(cx+d)$,

(iv) $(x+a)(x+b)/(x-a)(x-b)$,

(v) $a \cos x + b \sin x$,

(vi) $a \sec x + b \csc x$.

4.8. Approximation by polynomials. Taylor's theorem

The simplest class of functions with which a mathematician has to operate is the class of *polynomials*, that is to say,

$$a_0 + a_1 x + \ldots + a_n x^n,$$

where $a_0, \ldots, a_n$ are given numbers. The value of a polynomial is calculable exactly by additions and multiplications for any assigned value of the variable x.

More general functions are commonly expressible as limits of polynomials; in other words, the function can be expressed *approximately* by means of a polynomial. This section deals with the form that such an approximation normally takes.

Lemma. *If $f(x)$ is the polynomial*

$$a_0 + a_1 x + \ldots + a_n x^n,$$

then
$$a_r = \frac{1}{r!} f^{(r)}(0) \quad (0 \leqslant r \leqslant n).$$

Proof. The value of a_r is obtained by differentiating r times the polynomial expression for $f(x)$ and putting $x = 0$. ∎

Thus we have, if f is a polynomial of degree n,

$$f(x) = f(0) + xf'(0) + \frac{x^2}{2!} f''(0) + \ldots + \frac{x^n}{n!} f^{(n)}(0).$$

More generally, if we write $x = a + h$, where a is fixed, we have for the polynomial f

$$f(a+h) = f(a) + hf'(a) + \frac{h^2}{2!} f''(a) + \ldots + \frac{h^n}{n!} f^{(n)}(a).$$

We may reasonably expect that, if we discard the hypothesis that f is a polynomial but assume that it is a function possessing derivatives of the relevant orders, then the right-hand side of the last equation will be a good approximation to the value $f(a+h)$. This result is contained in theorems 4.81 and 4.82. They are seen to be forms of a mean value theorem of the nth order. The general mean value theorem is usually called Taylor's theorem after Brook Taylor (1685–1731) who investigated the expansion of a general function $f(a+h)$ in powers of h. The case $a = 0$ is commonly called Maclaurin's theorem.

In theorem 4.81 the existence of the nth derivative is required only for the single value a. The existence of $f^{(n)}(a)$ implies the existence of $f^{(n-1)}(x)$ in a neighbourhood of a; and this implies the continuity of $f^{(r)}(x)$ for $r \leqslant n-2$ in a neighbourhood of a.

Theorem 4.81. (*Young's form of the general mean value theorem*).
Suppose that $f^{(n)}(a)$ exists. Then

$$f(a+h) = f(a)+hf'(a)+\ldots+\frac{h^{n-1}}{(n-1)!}f^{(n-1)}(a)+\frac{h^n}{n!}\{f^{(n)}(a)+\epsilon\},$$

where $\epsilon \to 0$ as $h \to 0$.

Proof. Suppose first that $h \geqslant 0$. Define

$$\phi(h) = f(a+h)-f(a)-hf'(a)-\ldots$$

$$-\frac{h^{n-1}}{(n-1)!}f^{(n-1)}(a)-\frac{h^n}{n!}\{f^{(n)}(a)-\eta\},$$

where η is a positive constant.

Then $\phi(0), \phi'(0), \ldots, \phi^{(n-1)}(0)$ are all zero, and

$$\phi^{(n)}(0) = \eta.$$

By theorem 4.51, $\phi^{(n-1)}(h)$ is increasing at $h = 0$ and so is positive in an interval to the right of 0.

By corollary 2 to theorem 4.61, applied to $\phi^{(n-2)}(h)$, this function is positive in an interval to the right of 0.

Repeat the argument. We find finally that, for sufficiently small positive h, $\phi(h) > 0$, that is to say,

$$f(a+h) > f(a)+hf'(a)+\ldots+\frac{h^{n-1}}{(n-1)!}f^{(n-1)}(a)+\frac{h^n}{n!}\{f^{(n)}(a)-\eta\}.$$

Similarly

$$f(a+h) < f(a)+hf'(a)+\ldots+\frac{h^{n-1}}{(n-1)!}f^{(n-1)}(a)+\frac{h^n}{n!}\{f^{(n)}(a)+\eta\}.$$

A corresponding argument applies to negative values of h. Collecting our results, we have proved the theorem.

Application to maxima and minima. To illustrate the use of theorem 4.81 the reader should establish the following extension of the conditions given in theorem 4.73.

Suppose that

$$f'(c) = f''(c) = \ldots = f^{(n-1)}(c) = 0,$$

$$f^{(n)}(c) \neq 0.$$

Then (1) a necessary condition for c to give a maximum or minimum of $f(x)$ is that n is even; (2) supposing that n is even, if $f^{(n)}(c) < 0$, then c gives a maximum of $f(x)$, and if $f^{(n)}(c) > 0$, then c gives a minimum.

The next theorem is the direct extension to general n of theorem 4.61.

Theorem 4.82. (*Taylor's theorem.*) *Suppose that f and its derivatives up to order $n-1$ are continuous for $a \leqslant x \leqslant a+h$, and $f^{(n)}$ exists for $a < x < a+h$. Then*

$$f(a+h) = f(a) + hf'(a) + \ldots + \frac{h^{n-1}}{(n-1)!} f^{(n-1)}(a) + \frac{h^n}{n!} f^{(n)}(a+\theta h),$$

where $0 < \theta < 1$.

Proof. Define, for $0 \leqslant t \leqslant h$,

$$\phi(t) = f(a+t) - f(a) - tf'(a) - \ldots - \frac{t^{n-1}}{(n-1)!} f^{(n-1)}(a) - \frac{t^n}{n!} B,$$

where we choose B to make $\phi(h) = 0$.

From the definition of $\phi(t)$, we see that

$$\phi(0), \ \phi'(0), \ \phi''(0), \ \ldots, \ \phi^{(n-1)}(0)$$

are all zero. Now use Rolle's theorem (4.52) n times.

Since $\phi(0) = \phi(h) = 0$ we have $\phi'(h_1) = 0 \quad (0 < h_1 < h)$.

Since $\phi'(0) = \phi'(h_1) = 0$ we have $\phi''(h_2) = 0 \quad (0 < h_2 < h_1)$.

Finally,

since $\phi^{(n-1)}(0) = \phi^{(n-1)}(h_{n-1}) = 0$, we have $\phi^{(n)}(h_n) = 0$,

where $0 < h_n < h_{n-1} \ldots < h$ and so $h_n = \theta h \quad (0 < \theta < 1)$.

Now $\phi^{(n)}(t) = f^{(n)}(a+t) - B$, and so $B = f^{(n)}(a+\theta h)$.

Put $t = h$, $\phi(h) = 0$, and this value of B in the first line of the proof. |

A different proof of theorem 4.82 can be given, which has the advantage of yielding alternative expressions for R_n, the term in h^n. The form

$$R_n = \frac{h^n}{n!} f^{(n)}(a+\theta h)$$

is known as Lagrange's form of remainder.

For brevity we shall give this second proof in the Maclaurin form with $a = 0$.

6

Theorem 4.83. (*Taylor's theorem with Cauchy's form of remainder*). *With the hypotheses of theorem 4.82 (and $a = 0$),*

$$f(h) = f(0) + hf'(0) + \ldots + \frac{h^{n-1}}{(n-1)!} f^{(n-1)}(0) + R_n,$$

where $\qquad R_n = \dfrac{(1-\theta)^{n-1} f^{(n)}(\theta h) h^n}{(n-1)!} \quad (0 < \theta < 1).$

Proof. Define

$$F(t) = f(h) - f(t) - (h-t)f'(t) - \ldots - \frac{(h-t)^{n-1}}{(n-1)!} f^{(n-1)}(t).$$

Then it is easily verified that

$$F'(t) = -\frac{(h-t)^{n-1}}{(n-1)!} f^{(n)}(t),$$

all other terms in the differentiation cancelling in pairs.
Write

$$\Phi(t) = F(t) - \left(\frac{h-t}{h}\right)^p F(0),$$

where p can be any integer such that $1 \leqslant p \leqslant n$. Then

$$\Phi(0) = \Phi(h) = 0.$$

By Rolle's theorem (4.52), there is a value θh with $0 < \theta < 1$ for which $\Phi'(\theta h) = 0$. But

$$\Phi'(\theta h) = F'(\theta h) + \frac{p(1-\theta)^{p-1}}{h} F(0).$$

This gives a value for R_n which reduces to Lagrange's form when $p = n$ and to Cauchy's form stated in this theorem when $p = 1$. ∎

4.9. Indeterminate forms

We shall extend theorems 4.81 and 4.82 to two functions. (Compare theorem 4.62, when $n = 1$.)

Theorem 4.91. *Suppose that $f^{(n)}(a)$, $g^{(n)}(a)$ exist and $g^{(n)}(a) \neq 0$. Then, as $h \to 0$,*

$$\frac{f(a+h) - f(a) - hf'(a) - \ldots - \{h^{n-1}/(n-1)!\} f^{(n-1)}(a)}{g(a+h) - g(a) - hg'(a) - \ldots - \{h^{n-1}/(n-1)!\} g^{(n-1)}(a)} \to \frac{f^{(n)}(a)}{g^{(n)}(a)}.$$

Proof. Apply theorem 4.81 to each of the functions f, g. ∎

Theorem 4.92. *Suppose that f and g and their derivatives up to order $n-1$ are continuous for $a \leqslant x \leqslant a+h$. For $a < x < a+h$, suppose that $f^{(n)}(x)$ exists and that $g^{(n)}(x)$ exists and does not take the value 0. Then*

$$\frac{f(a+h)-f(a)-hf'(a)-\ldots-\{h^{n-1}/(n-1)!\}f^{(n-1)}(a)}{g(a+h)-g(a)-hg'(a)-\ldots-\{h^{n-1}/(n-1)!\}g^{(n-1)}(a)}$$
$$=\frac{f^{(n)}(a+\theta h)}{g^{(n)}(a+\theta h)},$$

where $0 < \theta < 1$.

Proof. We use the method of theorem 4.62. Define

$$\phi(x) = f(x) - kg(x).$$

Choose k to make

$$\phi(a+h)-\phi(a)-h\phi'(a)-\ldots-\frac{h^{n-1}}{(n-1)!}\phi^{(n-1)}(a)$$

equal to 0.

By theorem 4.82 the last expression is equal to

$$\frac{h^n}{n!}\phi^{(n)}(a+\theta h),$$

and therefore
$$f^{(n)}(a+\theta h)-kg^{(n)}(a+\theta h) = 0.$$

Equate the two values found for k. ∎

Corollary. *With the hypotheses of theorem 4.92, if*

$$f(a) = f'(a) = \ldots = f^{(n-1)}(a) = 0$$

and
$$g(a) = g'(a) = \ldots = g^{(n-1)}(a) = 0,$$

and
$$\frac{f^{(n)}(x)}{g^{(n)}(x)} \to l \quad as\ x \to a \quad (or\ a+\ or\ a-),$$

then
$$\frac{f(x)}{g(x)} \to l \quad as\ x \to a \quad (or\ a+\ or\ a-).$$

We now indicate a field of application of these theorems.

Suppose that f and g are continuous in the closed interval (a, b) and that $f(a) = g(a) = 0$. The limit as $x \to a+$ of the quotient $f(x)/g(x)$ cannot be ascertained directly by putting $x = a$. Such an expression is traditionally called an *indeterminate form* $(0/0)$. On suitable assumptions about the derivatives of f and g, the limit of $f(x)/g(x)$ can often be found by theorems 4.91 and 4.92.

It is to be observed that the hypotheses of theorems 4.91 and 4.92 are different, and neither theorem includes the other.

Illustration. Investigate the limit as $x \to 0$ of

$$\frac{\tan kx - k \tan x}{k \sin x - \sin kx}.$$

Solution. Writing the fraction as $f(x)/g(x)$, we have

$$f'(x) = k \sec^2 kx - k \sec^2 x,$$

$$g'(x) = k \cos x - k \cos kx.$$

If we use theorem 4.91 it is necessary to differentiate twice more, since g''' is the derivative of lowest order which does not vanish at $x = 0$.

Theorem 4.92 (corollary) is, however, applicable to give

$$\frac{f'(x)}{g'(x)} = \frac{\cos x + \cos kx}{\cos^2 kx \cos^2 x} \quad (x \neq 0),$$

and this tends to 2 as x to 0.

Exercises 4 (*f*)

Notes on these exercises are given on p. 177.

1. Find the limits, as x tends to $\frac{1}{2}$, of

$$\text{(i)} \quad \frac{(1-x)^m - x^m}{(1-x)^n - x^n},$$

where m and n are positive integers;

$$\text{(ii)} \quad \frac{1 + \cos 2\pi x}{(2x-1)^2}.$$

2. Find the equations of the tangent and normal to the ellipse $x = a \cos \theta$, $y = b \sin \theta$ at the point θ.

A tangent to an ellipse meets the axes at P and Q. Find the least value of PQ.

3. By writing $y = tx$ obtain parametric equations for the curve

$$x^3 + y^3 = 3axy.$$

Obtain the equation of the tangent at the point with parameter t, and find the parameter of the point where this tangent cuts the curve again. What happens as $t \to -1$?

4. Sketch the locus (the cycloid) given by

$$x = a(t - \sin t), \quad y = a(1 - \cos t)$$

for values of t between 0 and 2π.

Prove that the normals to the curve are tangents to the curve

$$x = a(t + \sin t), \quad y = -a(1 - \cos t),$$

and sketch this second curve in your diagram.

5. Investigate maxima and minima of

$$\text{(i)} \ \frac{|x-1|}{x^2+1}, \qquad \text{(ii)} \ \frac{\sin x}{\sin (x-a)},$$

and sketch the corresponding graphs.

6. Prove that, if $9b > a > 0$, the function

$$a \sin x + b \sin 3x$$

has a maximum for some value of x between 0 and $\frac{1}{2}\pi$. Taking $a = 6$, $b = 1$, find the greatest and least values of the function for $0 \leqslant x \leqslant \frac{1}{2}\pi$.

7. P is a point on the circle whose equation is

$$(x-h)^2 + (y-h)^2 = a^2,$$

where $h > a$, and PM, PN are the perpendiculars on the coordinate axes. Find the positions of P for which the area of the triangle PMN is a maximum, and show that there are two maximum positions or one according as h is less or greater than $a\sqrt{2}$.

8. If $$y = ax^3 + bx^2 + cx + d,$$

what condition must be satisfied by the coefficients a, b, c, d, if, corresponding to any value of y, there is only one value of x?

9. If $p(x)$ and $q(x)$ are quadratic, and the roots of $q(x) = 0$ are complex, prove that the maximum and minimum values of $p(x)/q(x)$ are the values of k for which $p(x) - kq(x)$ is a perfect square.

10. An open bowl is in the form of a segment of a sphere of metal of negligible thickness. Find the shape of the bowl if its volume is greatest for a given area of metal.

11. The centres of two spheres of radii a and b are at a distance c apart, where $c > a + b$. Where must a point source of light be placed on the line of centres between the two spheres so as to illuminate the greatest total surface?

12. A flat piece of cardboard has the form of an equilateral triangle ABC of height $3h$. Points P, Q, R are marked on the medians AG, BG, CG on opposite sides of G from A, B, C, with $GP = GQ = GR$. The triangles BCP, CAQ, ABR are then cut away, and the remaining piece of cardboard is folded about the edges of the triangle PQR so as to form the surface of a tetrahedron. Prove that the volume of this tetrahedron cannot exceed $(3/8)^{\frac{1}{2}}h^3$.

13. Construct a function f for which $f'(0) > 0$, but there is no interval $(-h, h)$ in which f is an increasing function. (Try $x^2 \sin (1/x) + kx$, where k is a suitable constant.)

14. Discuss *Newton's method of approximation*, that, if x_1 is near to a root a of the equation $f(x) = 0$, then

$$x_2 = x_1 - \frac{f(x_1)}{f'(x_1)}$$

is likely to be a better approximation to a.

Show in particular that if, in $(a-h, a+h)$, $|f'(x)| > k > 0$ and $f''(x)$ has constant sign, then x_2 lies between x_1 and a if $f(x_1)$ and $f''(a)$ have the same sign.

Approximate to the root near π of the equation $\sin x = \eta x$, where the constant η is small.

15. *The rule of proportional parts.* If the values of a function $f(x)$ are tabulated at intervals h, and $a+k$ lies between two successive entries a, $a+h$, the rule is that, approximately,

$$f(a+k) - f(a) = \frac{k}{h}\{f(a+h) - f(a)\}.$$

To obtain an upper bound for the error, suppose that $f''(x)$ exists for $a \leqslant x \leqslant a+h$. Write

$$\phi(t) = \frac{f(a+t) - f(a)}{t} \quad (0 < t < h, \ a \text{ fixed}).$$

Prove that $\qquad \phi'(t) = \tfrac{1}{2}f''(\tau) \quad (0 < \tau < t).$

Deduce that

$$\frac{f(a+h) - f(a)}{h} - \frac{f(a+k) - f(a)}{k} = \tfrac{1}{2}(h-k)f''(\kappa),$$

where $a < \kappa < a+h$.

Hence prove that the error in the value found by the rule of proportional parts is at most $\tfrac{1}{8}Mh^2$, where $M = \sup |f''(x)|$.

16. Prove that the nth derivative of $\tan x$ is a polynomial in $\tan x$ of the $(n+1)$th degree.

17. Prove that

$$\frac{d^n}{dx^n}\left(\frac{e^x}{x}\right) = (-1)^n n! \frac{e^x}{x^{n+1}} p_n(-x),$$

where $p_n(x)$ is the polynomial formed by the first $(n+1)$ terms of the expansion of e^x in powers of x.

18. Prove, by induction or otherwise, that, if m is a positive integer, $\sin(2m+1)\theta$ and $\cos(2m+1)\theta/\cos\theta$ can be expressed as polynomials in $\sin\theta$.

If $y = \sin(2m+1)\theta$ and $x = \sin\theta$, show that

$$(1-x^2)(y')^2 = (2m+1)^2(1-y^2)$$

and $\qquad (1-x^2)y^{(n+2)} - (2n+1)xy^{(n+1)} + \{(2m+1)^2 - n^2\}y^{(n)} = 0.$

Prove that

$$y = (2m+1)\left\{x - \frac{(2m+2)2m}{3!}x^3 + \frac{(2m+4)(2m+2)2m(2m-2)}{5!}x^5 - \ldots\right\},$$

19. A twice differentiable function is such that $f(a) = f(b) = 0$ and $f(c) > 0$, where $a < c < b$. Prove that there is at least one value ξ between a and b for which $f''(\xi) < 0$.

20. Prove that, as $h \to 0$,

$$\frac{f(a+h) - 2f(a) + f(a-h)}{h^2} \to f''(a)$$

if the right-hand side exists.

21. Investigate the limits as $x \to 0$ of

$$\text{(i)} \quad \frac{1}{x^2} - \frac{1}{\sin^2 x}, \qquad \text{(ii)} \quad \frac{\arcsin x - \sin x}{\tan x - \arctan x}.$$

22. Investigate the limit as $x \to 1$ of

$$\frac{x - (n+1)x^{n+1} + nx^{n+2}}{(1-x)^2}.$$

23. Find a pair of functions $f(x)$, $g(x)$ for which $f(0) = g(0) = 0$ and, as $x \to 0$, $f(x)/g(x)$ tends to a limit but $f'(x)/g'(x)$ does not tend to a limit. That is to say, prove that the converse of theorem 4.92, corollary (for $n = 1$) would be false. *Hint*—a discontinuous $f'(x)$ is given in exercise 4 (*b*), 10.

24. (*A theorem of Darboux.*) If $f(x)$ is differentiable for $a \leqslant x \leqslant b$, then $f'(x)$ takes every value between $f'(a)$ and $f'(b)$. (A derivative need not be continuous, but it has the property established in theorem 3.6.)

5

INFINITE SERIES

5.1. Series of positive terms

In chapter 2 we defined convergence of infinite series and gave some properties which followed quickly from the definition. Two particular series Σr^n and Σn^{-k} were studied. We now need a wider knowledge of infinite series.

We recall that a series of positive terms Σu_n either converges or diverges to $+\infty$. Throughout §5.1 we shall assume that $u_n \geqslant 0$.

The reader is asked to refer to the closing result (6) of §2.12. It enunciates a *comparison principle* by which it may be possible to infer the convergence of a proposed series Σu_n from that of a series Σv_n which is known to converge. We proceed to turn this comparison principle into readily applicable forms.

Theorem 5.11. (*Cauchy's test for convergence.*) *Suppose that* $u_n^{1/n}$ *tends to a limit* l *as* $n \to \infty$. *Then, if* $l < 1$, Σu_n *converges; if* $l > 1$, Σu_n *diverges.*

Proof. Suppose that $l < 1$. Choose r with $l < r < 1$.

By the definition of limit, there exists N such that, for all $n > N$,
$$u_n^{1/n} < r, \quad \text{i.e. } u_n < r^n.$$

But, since $r < 1$, the geometric series Σr^n converges. Therefore so does Σu_n.

If $l > 1$, then, for all n greater than some N,
$$u_n^{1/n} > 1, \quad \text{i.e. } u_n > 1$$
and so Σu_n diverges. ∎

Note carefully that, if $l = 1$, no conclusion can be drawn.

A criterion which is often easier to apply than theorem 5.11 is the following.

Theorem 5.12. (*d'Alembert's test*). *Suppose that* $u_n > 0$ *and* u_{n+1}/u_n *tends to* l. *Then, if* $l < 1$, Σu_n *converges; if* $l > 1$, Σu_n *diverges.*

Proof. Suppose that $l < 1$. Choose r with $l < r < 1$.

$$\exists\ N.\ \ \frac{u_{n+1}}{u_n} < r\ \ \text{for all } n \geqslant N.$$

Therefore

$$u_n = \frac{u_n}{u_{n-1}} \frac{u_{n-1}}{u_{n-2}} \cdots \frac{u_{N+1}}{u_N} u_N < u_N' r^{n-N},$$

i.e. $u_n < Kr^n$, where K is independent of n.

Since Σr^n converges, so does Σu_n.

If $l > 1$, the terms increase after a certain value N, and divergence is plain. $\blacksquare$

As in theorem 5.11, if $l = 1$, no conclusion can be drawn.

Exercises 5 (a)

Notes on these exercises are given on p. 177.

1. Investigate the convergence or divergence of the series whose nth terms are

$$\frac{2^{3n+1}}{3^{2n-1}}, \quad \frac{n^4}{2^n}, \quad 2^{-\sqrt{n}} x^n,$$

$$\frac{(n!)^2}{(2n)!} x^n, \quad \frac{1.3.5\ldots(2n-1)}{1.4.7\ldots(3n-2)} x^n,$$

where $x > 0$.

2. Prove that the series whose nth term is

$$\frac{(a+1)(2a+1)\ldots(na+1)}{(b+1)(2b+1)\ldots(nb+1)}$$

converges if $0 < a < b$ and diverges if $a \geqslant b > 0$.

3. Prove that, if $u_n^{1/n} \leqslant r < 1$ for all $n > N$, then Σu_n converges.

Show that this statement provides a test which is more general than theorem 5.11.

Discuss the series

$$a + b + a^2 + b^2 + a^3 + b^3 + \ldots,$$

where $0 < a < b < 1$.

4. State a test for convergence which is more general than d'Alembert's test in the same way that exercise **3** is more general than Cauchy's test.

5. Prove that d'Alembert's test does not determine the convergence of the series in exercise **3**.

6. Discuss the following statement:

The tests of Cauchy and d'Alembert, being derived from comparison with a geometric progression, cannot determine the convergence of the series Σn^{-2}, which converges more slowly than any geometric progression (§2.11, exercise).

7. Prove that if, for all n, u_{n+3}/u_n is less than k, where $k < 1$, then Σu_n converges. Generalise.

8. Establish the truth or falsity of these statements:

   (i) If $(u_n/v_n) \to 1$ as $n \to \infty$, then Σu_n and Σv_n both converge or both diverge.

   (ii) If $u_n - v_n \to 0$, then Σu_n and Σv_n both converge or both diverge.

   (iii) If $(u_{n+1}/u_n) > k > 1$ for infinitely many n, then Σu_n diverges.

5.2. Series of positive and negative terms

We turn to the problem of settling the convergence or divergence of a series which has infinitely many positive and infinitely many negative terms (for example, the series $1 - \frac{1}{2} + \frac{1}{3} - \frac{1}{4} + \ldots$). The criteria of theorems 5.11 and 5.12, applicable to series of positive terms, depended on inequalities between the terms of the series under investigation and those of a series whose behaviour was known. A little thought shows that no such 'comparison principle' holds for series of terms of arbitrary sign.

We ask, then, whether our knowledge of series of positive terms can be turned to account. If we are given any series Σu_n, there is one series of positive terms which suggests itself as being closely related to it, namely, the series of absolute values of the terms, $\Sigma |u_n|$. This observation leads to a satisfying theorem.

Theorem 5.21. *If $\Sigma |u_n|$ converges, then Σu_n converges.*

   Proof. Define

$$v_n = \begin{cases} u_n & \text{if } u_n \geqslant 0, \\ 0 & \text{if } u_n < 0. \end{cases}$$

$$w_n = \begin{cases} 0 & \text{if } u_n \geqslant 0, \\ -u_n & \text{if } u_n < 0. \end{cases}$$

Then $v_n \geqslant 0$, $w_n \geqslant 0$ and the series Σv_n contains those terms of Σu_n which are positive (or zero).

   Also

$$u_n = v_n - w_n,$$

$$|u_n| = v_n + w_n.$$

If now $\Sigma |u_n|$ converges, then Σv_n and Σw_n both converge. Therefore so does $\Sigma(v_n - w_n)$, i.e. Σu_n. ∎

Definition. *If $\Sigma|u_n|$ is convergent, the series Σu_n is called* absolutely convergent.

Exercise. Give three examples of series which converge, but not absolutely. (An example is $1-1+\frac{1}{2}-\frac{1}{2}+\frac{1}{3}-\frac{1}{3}+\dots$. Theorem 5.22 will suggest many others.)

If we have to determine whether a series Σu_n converges, the first step is to look at $\Sigma|u_n|$. If $\Sigma|u_n|$ converges, the matter is settled. If $\Sigma|u_n|$ diverges we have to try other methods.

The most common distribution of signs in series is $+$ and $-$ alternately. The following is a very useful theorem on such 'alternating series'.

Theorem 5.22. *If a_n decreases and tends to zero as $n \to \infty$, then the series*

$$a_0 - a_1 + a_2 - a_3 + \dots$$

converges. Also its sum lies between a_0 and $a_0 - a_1$.

　　Proof. If
$$s_n = a_0 - a_1 + \dots + (-1)^n a_n,$$

then
$$s_{2n+1} - s_{2n-1} = a_{2n} - a_{2n+1} \geqslant 0,$$

$$s_{2n} - s_{2n-2} = -a_{2n-1} + a_{2n} \leqslant 0.$$

The sums with even suffix

$$s_0, \; s_2, \; s_4, \; \dots$$

thus form a decreasing sequence. By theorem 2.6 this tends to a limit or to $-\infty$.

Similarly the sequence of sums with odd suffix

$$s_1, \; s_3, \; s_5, \; \dots$$

increases and tends to a limit or to $+\infty$. But

$$s_{2n+1} - s_{2n} = -a_{2n+1} \to 0.$$

Therefore the odd and even sequences must have the same finite limit, and so s_n tends to this limit.

Finally, $s_0 = a_0$ and $s_1 = a_0 - a_1$. So the sum of the series lies between these two numbers. ▮

Illustration. For what values of k is the series

$$\frac{1}{1^k} - \frac{1}{2^k} + \frac{1}{3^k} - \frac{1}{4^k} + \dots$$

convergent?

If $k > 1$, the series is absolutely convergent (theorem 2.11).

If $0 < k \leqslant 1$, it converges (theorem 5.22). Note that it is *not* absolutely convergent.

If $k \leqslant 0$, the nth term does not tend to zero, and the series diverges.

Exercises 5 (*b*)

Notes on these exercises are given on pp. 177–8.

1. Σa_n is a convergent series of positive terms. Prove that

    (i) if $|b_n| \leqslant a_n$, then Σb_n is absolutely convergent;

    (ii) $\Sigma a_n x^n$ is absolutely convergent for $-1 \leqslant x \leqslant 1$;

    (iii) $\Sigma a_n \cos n\theta$ and $\Sigma a_n \sin n\theta$ are absolutely convergent.

2. For what values of k is $\Sigma(-1)^n/(2n+1)^k$ absolutely convergent? For what values of k is it convergent?

3. For what values of x (if any) are the following series convergent but not absolutely convergent:

$$\text{(i) } \Sigma x^n, \quad \text{(ii) } \Sigma x^n/n?$$

4. Discuss the convergence of the series:

$$\text{(i) } 1 - \tfrac{3}{8} + \tfrac{5}{16} - \tfrac{7}{24} + \tfrac{9}{32}\ldots;$$

$$\text{(ii) } 1 - \tfrac{3}{2} + \tfrac{5}{4} - \tfrac{7}{8} + \tfrac{9}{16} - \ldots;$$

$$\text{(iii) } 1 - \tfrac{3}{1} + \tfrac{5}{4} - \tfrac{7}{9} + \tfrac{9}{16} - \ldots,$$

the respective denominators being of the form $8n$, 2^n and n^2.

5. Discuss the convergence of the series

$$\frac{a}{1} - \frac{b}{2} + \frac{a}{3} - \frac{b}{4} + \ldots + \frac{a}{2m-1} - \frac{b}{2m} + \ldots,$$

where a and b are positive constants.

6. Find two numbers differing by not more than $\tfrac{1}{6}$ between which the sum of the series

$$1 - \tfrac{1}{2} + \tfrac{1}{3} - \tfrac{1}{4} + \tfrac{1}{5} - \tfrac{1}{6} + \ldots$$

must lie.

7. Establish the truth or falsity of the following statements:

    (i) If $a_n > 0$ and a_n tends to 0 as $n \to \infty$, then the series

$$a_0 - a_1 + a_2 - a_3 + \ldots$$

converges.

    (ii) If Σv_n is absolutely convergent, and there is a constant A such that

$$-A|v_n| \leqslant u_n \leqslant A|v_n| \quad \text{for all } n,$$

then Σu_n is absolutely convergent.

5.3. Conditional convergence

Definition. *If Σu_n converges and $\Sigma |u_n|$ diverges, then Σu_n is said to be conditionally convergent.*

Illustration. The series
$$1-\tfrac{1}{2}+\tfrac{1}{3}-\tfrac{1}{4}+\ldots$$
converges conditionally.

The meaning of the word conditional here is that the sum to which the series converges is conditional on the order in which the terms are taken. If the terms are rearranged the sum is in general altered. As an illustration, rearrange the last series so that two negative terms always follow one positive term

$$1-\tfrac{1}{2}-\tfrac{1}{4}+\tfrac{1}{3}-\tfrac{1}{6}-\tfrac{1}{8}+\ldots.$$

Suppose that the sum of the original series is s (in fact, $s = \log_e 2$ but we do not need to know this). We shall prove that the rearranged series converges to sum $\tfrac{1}{2}s$. Let s_n and t_n be the sums of the first n terms of the two series. Then

$$t_{3n} = 1 - \frac{1}{2} - \frac{1}{4} + \ldots + \frac{1}{2n-1} - \frac{1}{4n-2} - \frac{1}{4n}.$$

In each block of three terms (two negative and one positive) subtract the first of the negative terms from the preceding positive term, and we have

$$t_{3n} = \frac{1}{2} - \frac{1}{4} + \frac{1}{6} - \frac{1}{8} + \ldots + \frac{1}{4n-2} - \frac{1}{4n}$$
$$= \tfrac{1}{2}s_{2n}.$$

Therefore, as $n \to \infty$, $t_{3n} \to \tfrac{1}{2}s$; t_{3n+1} and t_{3n+2} tend to the same limit. So the rearranged series converges to sum $\tfrac{1}{2}s$.

This change of sum by rearrangement is not paradoxical, as s_n and t_n are different functions of n. The more striking fact is that, however we alter the order of the terms of an absolutely convergent series, the sum is unchanged.

Theorem 5.3. *If Σu_n is absolutely convergent, every series consisting of the same terms in any order has the same sum.*

Proof. We prove the theorem first in the special case in which all the terms are positive ($u_n \geqslant 0$).

Let $\Sigma u_n'$ consist of the same terms as Σu_n with the order rearranged in any way.

Let
$$s_m = \sum_1^m u_n, \quad s = \sum_1^\infty u_n, \quad t_m = \sum_1^m u_n'.$$

Every term of $\Sigma u'_n$ occurs somewhere in Σu_n. Therefore, given m, we can find q such that s_q contains every term of t_m. Since the terms are positive, $t_m \leqslant s_q \leqslant s$. As $m \to \infty$, t_m tends to a limit t where $t \leqslant s$. We can now argue the other way round and prove $s \leqslant t$. This establishes the result for series of positive terms. |

The theorem must now be proved when Σu_n is any absolutely convergent series and $\Sigma u'_n$ a rearrangement of it.

As in theorem 5.21 define v_n and $-w_n$ to be the positive and negative terms of Σu_n; and v'_n, w'_n to have the same significance for the series $\Sigma u'_n$.

Since $\Sigma |u_n|$ converges, Σv_n and Σw_n both converge.

Then $\Sigma v'_n$ and $\Sigma w'_n$ are rearrangements respectively of the convergent series of positive terms Σv_n and Σw_n.

The result follows from the special case of the theorem. |

Exercise. If s_n is the sum of the first n terms of the series

$$1 - \frac{1}{\sqrt{2}} + \frac{1}{\sqrt{3}} - \frac{1}{\sqrt{4}} + \dots$$

and t_n is the sum of the first n terms of the series rearranged by taking two positive terms followed by one negative term

$$1 + \frac{1}{\sqrt{3}} - \frac{1}{\sqrt{2}} + \frac{1}{\sqrt{5}} + \frac{1}{\sqrt{7}} - \frac{1}{\sqrt{4}} + \dots,$$

prove that
$$t_{3n} > s_{2n} + \frac{n}{\sqrt{(4n-1)}}$$

and deduce that the second series diverges to $+\infty$.

5.4. Series of complex terms

The notions of limit of a sequence and convergence of a series can be extended from real to complex numbers.

Let
$$s_n + it_n = \sum_{r=1}^{n} (u_r + iv_r).$$

Definition. We say that $s_n + it_n \to s + it$ and $\Sigma(u_n + iv_n)$ converges to sum $s + it$ if

$$|(s+it) - (s_n + it_n)| \to 0 \quad \text{as} \quad n \to \infty.$$

This is equivalent to saying that

$$s_n \to s \quad \text{and} \quad t_n \to t,$$

because
$$|s - s_n| \leqslant |(s - s_n) + i(t - t_n)| \leqslant |s - s_n| + |t - t_n|.$$

Definition. We say that the series $\Sigma(u_n + iv_n)$ is *absolutely convergent* if $\Sigma|u_n + iv_n|$ is convergent.

This is equivalent to saying that both Σu_n and Σv_n are absolutely convergent for (as above),

$$|u_n| \leqslant |u_n + iv_n| \leqslant |u_n| + |v_n|.$$

Many results for real sequences and series can be extended to complex sequences and series, and the proofs offer no difficulty. We need, for instance, the following analogue of §2.7.

The sequence z^n (for fixed z) tends to a limit as $n \to \infty$ if and only if $z = 1$ or $|z| < 1$.

Proof. Suppose that $z^n \to l$. Then $z^{n+1} \to l$. But the limit of z^{n+1}, i.e. of $(z)(z^n)$ is the product of z and the limit of z^n, i.e. zl. So
$$l = zl.$$

This is true if and only if either $z = 1$ or $l = 0$.

But $z^n \to 0$ if and only if $|z|^n \to 0$, i.e. (from §2.6) if and only if $|z| < 1$. |

This knowledge of the behaviour of z^n enables us to discuss the convergence of the geometric series

$$1 + z + \ldots + z^n + \ldots.$$

If $z = 1$, the series diverges.

If $z \neq 1$, $s_n = (1 - z^n)/(1 - z)$ and we have seen that this tends to a limit if and only if $|z| < 1$.

So the values of z for which the series converges are the points inside a circle in the complex plane. We shall see in the next section that this 'circle of convergence' exists for a wide class of series.

Exercises 5 (c)

Notes on these exercises are given on p. 178.

1. Find the sum to n terms of the series
$$1 + 2z + 3z^2 + \ldots + (n+1)z^n + \ldots.$$
Prove that, if $|z| < 1$, the series converges and find its sum.

2. Prove that the series

$$1 - \frac{z}{1-z} + \left(\frac{z}{1-z}\right)^2 - \left(\frac{z}{1-z}\right)^3 + \dots$$

converges if and only if re $z < \frac{1}{2}$. What is its sum?

3. Decide the convergence or divergence of the series whose nth terms are

$$\text{(i)} \ \frac{i^n}{n}, \quad \text{(ii)} \ \frac{(1+i)^n}{n^2}, \quad \text{(iii)} \ \frac{i^{n^2}}{n}.$$

5.5. Power series

A series $\sum\limits_{n=0}^{\infty} a_n z^n$ of multiples of powers of z is called a *power series*. In practice the variable z and the coefficients a_n are often real, but we can discuss the series with nearly as great ease if they may have complex values.

Theorem 5.51. *A power series may converge* (1) *for all values of z, or* (2) *for z in some region in the complex plane, or* (3) *for $z = 0$ only.*

Proof. All we have to do is to produce examples of each possibility.

(1) If $u_n = z^n/n!$, Σu_n is absolutely convergent for all values of z. For, whatever the value of z,

$$\frac{|u_{n+1}|}{|u_n|} = \frac{|z|}{n+1} \to 0$$

and d'Alembert's test (theorem 5.12) gives the result.

(2) The geometric series Σz^n was proved in §5.4 to converge if and only if $|z| < 1$.

(3) If $u_n = n! \, z^n$, and $z \neq 0$, $|u_n| \to \infty$ as $n \to \infty$ and Σu_n cannot converge.

Theorem 5.52. *If a power series converges for a particular value of z, say $z = z_1$, then it converges absolutely for all values of z in the circle $|z| < |z_1|$.*

Proof. Since $\Sigma a_n z_1^n$ converges, therefore the nth term $a_n z_1^n$ tends to 0 (§2.12 (3)). So we can find K such that $|a_n z_1^n| < K$ for all n. Then

$$|a_n z^n| < K \left|\frac{z}{z_1}\right|^n$$

and the convergence of $\Sigma |a_n z^n|$ follows from that of the geometric series $\Sigma |z/z_1|^n$.

Exercise. Show that the conclusion of theorem 5.52 is true under the wider hypothesis that $\Sigma a_n z^n$ oscillates finitely for $z = z_1$.

5.6. The circle of convergence of a power series

Theorem 5.61. *A power series either*

(1) *converges absolutely for all z, or*

(2) *converges absolutely for all z inside a circle $|z| = R$ and diverges for all z outside it, or*

(3) *converges for $z = 0$ only.*

Proof. Let x be a positive real number ($x \geqslant 0$).

Let S be the set of x for which the power series $\Sigma a_n x^n$ converges. $x = 0$ is certainly in S; from theorem 5.51, S may or may not contain members other than 0. By theorem 5.52, if any x_1 is in S, so is every x with $0 \leqslant x \leqslant x_1$.

If all positive real numbers are in S we have the case (1) of the conclusion.

If S does not contain all the positive numbers it has a finite supremum R (where $R \geqslant 0$).

If $R > 0$, we shall prove that $\Sigma a_n z_1^n$ converges absolutely if $|z_1| < R$. For choose R_0 with $|z_1| < R_0 < R$. Then R_0 is in S and so the series converges for $z = R_0$. By theorem 5.52, $\Sigma |a_n z_1^n|$ converges.

Next we prove that, if $|z_2| > R \geqslant 0$, the series cannot converge for $z = z_2$. For take now R_0 with $R < R_0 < |z_2|$. If $\Sigma a_n z_2^n$ were to converge, then, by theorem 5.52, $\Sigma a_n R_0^n$ would converge, which contradicts $R = \sup S$. |

Definitions. *The circle $|z| = R$ is called the* circle of convergence *of the power series and its radius the* radius of convergence.

It is to be noted that nothing has been proved about convergence or divergence of the series for values of z *on* the circle of convergence. This is more delicate and requires special investigation for any particular series.

The following simple formula for R applies to many common series.

Theorem 5.62. *If* $|a_{n+1}/a_n|$ *tends to a limit* l *as* $n \to \infty$, *then the radius of convergence of* $\Sigma a_n z^n$ *is* $1/l$.

Proof. By d'Alembert's test we have absolute convergence if

$$\lim \left| \frac{a_{n+1} z^{n+1}}{a_n z^n} \right| < 1,$$

i.e. if
$$|z| < \frac{1}{l}.$$

And, if $|z| > 1/l$,
$$\lim \left| \frac{a_{n+1} z^{n+1}}{a_n z^n} \right| > 1,$$

so that the nth term $a_n z^n$ does not tend to zero and convergence is impossible. ∎

Note. We have supposed l finite and not zero. The reader should consider the excepted cases.

Illustrations. Each of the series

$$\Sigma z^n, \quad \Sigma(z^n/n), \quad \Sigma(z^n/n^2)$$

has radius of convergence 1. The first series converges at no point on the circle $|z| = 1$; the third is absolutely convergent at all points of the circle. The second diverges for $z = 1$, and it can be shown to converge at all other points on the circle $|z| = 1$ by the device suggested in exercise 5 (d), 8. The systematic treatment of the question of convergence of $\Sigma a_n z^n$ on its circle of convergence is outside the scope of this book.

Exercises 5 (d)

Notes on these exercises are given on p. 178.

1. Find the radii of convergence of the power series of which the general terms are

(i) nz^n,
(ii) $n^3 z^n/n!$
(iii) $\left(\dfrac{nz}{n+1} \right)^n$,

(iv) $n! z^n$,
(v) $\dfrac{(n!) z^{2n}}{(2n)!}$,
(vi) $\dfrac{z^{2n+1}}{2n+1}$,

(vii) $n! z^{n^2}$,
(viii) $\{3 + (-1)^n\}^n z^n$.

2. Prove that, if $|a_n|^{1/n} \to 1/r$ as $n \to \infty$, then the series $\Sigma a_n z^n$ has radius of convergence r.

Discuss the convergence of the series

$$\Sigma a^{n^2} z^n$$

where a is a constant.

3. If the series $\Sigma a_n z^n$ has radius of convergence R, in what region of the z-plane does the series $\Sigma a_n(z - z_0)^n$ converge? Answer the same question for the series $\Sigma a_n z^{-n}$.

4. Prove that the series
$$1+z+z^2+\ldots+z^n+\ldots$$
converges for all values of z for which the series
$$2+2(2z-1)+2(2z-1)^2+\ldots+2(2z-1)^n+\ldots$$
converges, and has the same sum.

5. If, for all n, $|a_n| < k$, what can you say about the radius of convergence of $\Sigma a_n z^n$? If, further, $|a_n| > l > 0$, what then follows?

6. If, for all n, $|a_n| \leqslant 1$, then the equation
$$1 = a_1 z + a_2 z^2 + \ldots$$
cannot have a root with modulus less than $\frac{1}{2}$. If it is satisfied by
$$z = \tfrac{1}{2}(\cos\theta + i\sin\theta),$$
then
$$a_n = \cos n\theta - i\sin n\theta.$$

7. If the radius of convergence of $\Sigma a_n z^n$ is r and of $\Sigma b_n z^n$ is s, what can you say about the radius of convergence of
$$\Sigma(a_n+b_n)z^n?$$

8. By considering $(1-z)\sum_1^m (z^n/n)$ when $|z| = 1$ or otherwise, prove that $\sum_1^\infty (z^n/n)$ converges at all points of the circle $|z| = 1$ except $z = 1$.

9. Discuss the convergence of the power series whose nth term is
$$\frac{1.3.5\ldots(2n-1)}{1.4.7\ldots(3n-2)}\,z^n.$$

5.7. Multiplication of series

We may wish to multiply two infinite series together and to know whether it is legitimate to write, say, the product of two power series
$$(a_0+a_1 z+a_2 z^2+\ldots)\,(b_0+b_1 z+b_2 z^2+\ldots)$$
as the power series
$$a_0 b_0 + (a_1 b_0 + a_0 b_1)\,z + (a_2 b_0 + a_1 b_1 + a_0 b_2)\,z^2 + \ldots.$$

This is the extension to infinite series of the product of two polynomials.

Since the process of multiplication involves freedom to arrange the terms of the product in the required order, we may conjecture that a sufficient condition for its validity will be the absolute convergence of the two series. We shall prove the correctness of this conjecture. The theorem will be stated for series which need not be power series.

Theorem 5.7. *If Σu_n and Σv_n converge absolutely to sums s and t, then the series $\Sigma u_p v_q$, consisting of the products (in any order) of every term of the first series by every term of the second, converges absolutely to sum st.*

Proof. The products of pairs of terms form a doubly infinite array

$$
\begin{vmatrix}
u_0 v_0 & u_0 v_1 & u_0 v_2 & \cdots \\
u_1 v_0 & u_1 v_1 & u_1 v_2 & \cdots \\
u_2 v_0 & u_2 v_1 & u_2 v_2 & \cdots \\
\cdots & \cdots & \cdots & \cdots
\end{vmatrix}.
$$

The sum of all these terms can be arranged (in infinitely many ways) as a single series. For example, we may take the terms $u_p v_q$, where $p + q = n$ in the order of increasing n, namely

$$u_0 v_0 + (u_1 v_0 + u_0 v_1) + (u_2 v_0 + u_1 v_1 + u_0 v_2) + \cdots.$$

This may be called diagonal summation. Or we may 'sum by squares', as in

$$u_0 v_0 + (u_1 v_0 + u_1 v_1 + u_0 v_1) + (u_2 v_0 + u_2 v_1 + u_2 v_2 + u_2 v_1 + u_2 v_0) + \cdots,$$

taking $u_0 v_0$, then terms with a suffix 1 and no greater, then those with a suffix 2 and no greater, and so on.

Whatever the arrangement, the sum of the moduli of any number of terms of the product does not exceed

$$\left(\sum_0^\infty |u_n| \right) \left(\sum_0^\infty |v_n| \right),$$

and so the series $\Sigma u_p v_q$ converges absolutely.

By theorem 5.3 (which remains true for complex terms), its sum is the same whatever the order of the terms. But there is one particular order, namely, summation by squares, in which the sum is evident. For the sum of all terms with suffixes not exceeding n is

$$(u_0 + u_1 + \ldots + u_n)(v_0 + v_1 + \ldots + v_n)$$

and the limit of this is st. ∎

Corollary. *If* $\Sigma a_n z^n$ *and* $\Sigma b_n z^n$ *have radii of convergence* R *and* S, *then their product is*

$$\sum_{n=0}^{\infty} (a_n b_0 + a_{n-1} b_1 + \dots + a_0 b_n)\, z^n$$

for $|z| < \min(R, S)$.

5.8. Taylor's series

A theorem of fundamental importance states that, if $f(x)$ satisfies certain conditions, then it can be expanded in the power series

$$f(x) = f(0) + x f'(0) + \dots + \frac{x^n}{n!} f^{(n)}(0) + \dots.$$

We proceed to prove this.

As the notation indicates, the variable is real. There is indeed an analogous expansion of $f(z)$, which is even more far-reaching than the expansion of $f(x)$, but it belongs to a later stage in your mathematical education.

Where has anything like this expression for $f(x)$ occurred in this book so far? Theorem 4.82, a mean value theorem of the nth order, expresses $f(x)$ as a polynomial in x, of which the coefficients up to that of x^{n-1} are those of the power series. The passage from the polynomial to the infinite series is valid under the hypotheses of the following theorem.

Theorem 5.8. (*Taylor's series*). *Suppose that* $f(x)$ *has derivatives of every order for* $a - k < x < a + k$. *Suppose also that, if* h *is any number such that* $|h| < k$ *and* θ *any number such that* $0 < \theta < 1$, *then*

$$\frac{h^n}{n!} f^{(n)}(a + \theta h) \to 0 \quad as \quad n \to \infty.$$

Then, if $|h| < k$,

$$f(a + h) = f(a) + h f'(a) + \dots + \frac{h^n}{n!} f^{(n)}(a) + \dots.$$

Proof. Let
$$S_n = \sum_{r=0}^{n-1} \frac{h^r}{r!} f^{(r)}(a)$$

and
$$R_n = \frac{h^n}{n!} f^{(n)}(a + \theta h).$$

From theorem 4.82,

$$f(a+h) = S_n + R_n,$$

where θ satisfies $0 < \theta < 1$.

Since $R_n \to 0$, we have $S_n \to f(a+h)$ as $n \to \infty$. ∎

As we remarked on p. 79, the expansion for $a = 0$ is often named after Maclaurin.

The binomial series. As an example of the use of theorem 5.8, we shall prove the binomial theorem for $f(x) = (1+x)^m$, where m is a rational number, positive or negative.

$$f^{(n)}(x) = m(m-1) \ldots (m-n+1)(1+x)^{m-n}$$

and we shall prove that, if $-1 < x < 1$,

$$(1+x)^m = 1 + \binom{m}{1}x + \ldots + \binom{m}{n}x^n + \ldots.$$

If m is a positive integer, $f^{(m+1)}(x) \equiv 0$ and we have a polynomial of degree m. In the general case

$$R_n = \frac{x^n}{n!}f^{(n)}(\theta x) = \binom{m}{n}\frac{x^n}{(1+\theta x)^{n-m}},$$

where θ depends on n.

If now $0 \leqslant x < 1$, $(1+\theta x)^{n-m} > 1$ for $n > m$, and, as stated in exercise 2(d), 11,

$$\binom{m}{n}x^n \to 0 \quad \text{as} \quad n \to \infty,$$

so that $R_n \to 0$.

If x is negative, this argument breaks down since $1+\theta x$ is not greater than 1. We have recourse to Cauchy's form of remainder (theorem 4.83) which applies for the full range $-1 < x < 1$. This gives

$$R_n = \frac{m(m-1)\ldots(m-n+1)}{1.2\ldots(n-1)}\frac{(1-\theta)^{n-1}x^n}{(1+\theta x)^{n-m}}.$$

Now $(1-\theta)/(1+\theta x)$ is less than 1 and so

$$|R_n| < K_m \left|\binom{m-1}{n-1}\right| |x|^n,$$

where K_m depends on m (and x) but not on n. Again from exercise 2(d), 11, $R_n \to 0$ as $n \to \infty$. ∎

Other well-known Taylor's series will be found in the next chapter.

Exercises 5 (e)

Notes on these exercises are given on p. 178.

1. Discuss the convergence of the series whose nth terms are (with a, b, c positive)

   (i) $1 - \cos(\pi/n)$,   (ii) $a^n/(b^n + c^n)$,

   (iii) $\dfrac{1}{(an^2 + bn + c)^k}$,   (iv) $\dfrac{(-1)^n}{(an^2 + bn + c)^k}$,

   (v) $(-1)^n a^{1/n}$ $(a > 0)$,   (vi) $(-1)^n \{\sqrt{(n^2 + 1)} - n\}$.

2. Prove that, if $a > 1$, the series

$$\frac{1}{a+1} + \frac{2}{a^2+1} + \frac{4}{a^4+1} + \frac{8}{a^8+1} + \cdots$$

converges to the sum $1/(a-1)$.

3. If $u_n \geqslant u_{n+1}$ and Σu_n converges, prove that $\lim n u_n = 0$.

4. From the series $\Sigma(1/n)$, every term which contains a specified digit (say 7) is removed. Prove that the series formed by the remaining terms converges.

5. The series $\Sigma a_n z^n$ converges to sum $f(z)$ if $|z| < 1$. Prove that, if $s_n = a_0 + a_1 + \ldots + a_n$ and $|z| < 1$, the series $\Sigma s_n z^n$ converges] to sum $f(z)/(1-z)$.

Hence give a proof of the binomial theorem when the exponent is a negative integer.

6

THE SPECIAL FUNCTIONS
OF ANALYSIS

6.1. The special functions of analysis

One of the ultimate applications of mathematical analysis is to solve in a form adapted to numerical calculation the problems which present themselves in natural science, engineering, economics and other branches of knowledge. Commonly the step from the experimental data or the hypotheses to the conclusions that they can be made to yield lies in solving *differential equations*. Relations are given connecting an unknown function with its first or second (or higher) derivatives, and the function has to be found. The analyst is led to keep a stock of such functions as occur repeatedly. He will investigate their properties, tabulate their numerical values, and have them ready for use. Such functions may be called the *special functions of analysis*. The list of special functions is not fixed, once and for ever. One mathematician might suppose some particular function to be of so little interest that he would not accord it a place in a list or think that the labour of tabulating its values would be justified. Another might encounter problems in which just that function played a leading part. There are, however, certain functions which are of vital importance to every one. Among them are the exponential, logarithmic and trigonometric functions; we shall develop their principal properties.

The functions which arise first from the foundations of analysis are those which are generated by a finite number of operations on the variable x. Such operations yield successively the function x^n, polynomials in x, and then rational functions of x. To obtain functions other than rational functions, we must remove the restriction to a finite number of operations, or, in other words, we must admit limiting processes. We can then expect to define interesting functions as the sums of infinite series.

6.2. The exponential function

Define

$$\exp x = 1 + x + \frac{x^2}{2!} + \ldots + \frac{x^n}{n!} + \ldots.$$

From theorem 5.51, the series converges for all values of x, real or complex. We shall suppose, until further notice, that x is real. The function $\exp x$ will be proved to have properties of striking simplicity.

Theorem 6.2. $\exp x \times \exp y = \exp (x+y)$.

Proof. If we use theorem 5.7 (corollary) to multiply the two series for $\exp x$ and $\exp y$, the terms of degree n in x and y are

$$\frac{x^n}{n!} + \ldots + \frac{x^r}{r!} \frac{y^{n-r}}{(n-r)!} + \ldots + \frac{y^n}{n!} = \frac{(x+y)^n}{n!}. \ \blacksquare$$

The following facts are immediate.

$\exp 0 = 1$.

$\exp (-x) = 1/\exp x$. (Put $y = -x$ in theorem 6.2.)

$\exp x$ never vanishes. (For then, from the last line, $\exp (-x)$ would be undefined.)

We shall next prove that

$$\frac{d}{dx} (\exp x) = \exp x.$$

We see that if we write down the derivatives of the successive terms of the series for $\exp x$, they are indeed the terms of $\exp x$. So the result *looks* right, but it is important that the reader shall understand why care is necessary here in constructing a proof. The next section on *repeated limits* is inserted to explain the issue.

6.3. Repeated limits

We know from §4.2 that the derivative of the sum of a *finite* number of functions is the sum of the derivatives of the separate functions. But we have proved no such theorem about the derivative of the sum of an infinite series; and this is the result we should need in order to deduce that the derivative of $\exp x$ is $\exp x$.

To see what is involved, write $s_n(x)$ for the sum of the first n terms of a convergent infinite series, whose terms are functions of x, and $s(x)$ for its sum. Then $s(x) = \lim\limits_{n \to \infty} s_n(x)$. We wish to assert that

$$s'(x) = \lim_{n \to \infty} s_n'(x),$$

i.e. $\quad \lim\limits_{h \to 0} \dfrac{s(x+h) - s(x)}{h} = \lim\limits_{n \to \infty} \left\{ \lim\limits_{h \to 0} \dfrac{s_n(x+h) - s_n(x)}{h} \right\},$

i.e. $\quad \lim\limits_{h \to 0} \left\{ \lim\limits_{n \to \infty} \dfrac{s_n(x+h) - s_n(x)}{h} \right\} = \lim\limits_{n \to \infty} \left\{ \lim\limits_{h \to 0} \dfrac{s_n(x+h) - s_n(x)}{h} \right\}.$

So we have got down to the root of the matter. The truth of the theorem depends on interchanging the order of the two limiting operations

$$\lim_{h \to 0} \quad \text{and} \quad \lim_{n \to \infty}$$

applied to the particular function of n and h (and also of x, which remains fixed while n and h vary).

All we can say is that the interchange of order of two limiting operations in general gives different results. It is only when the function to which they are applied satisfies restrictive conditions that the interchange is valid. A simple illustration in which the order of the two limits affects the result is the following:

$$\lim_{h \to 0} \left\{ \lim_{n \to \infty} \frac{1 - nh}{1 + nh} \right\} = \lim_{h \to 0} (-1) = -1,$$

$$\lim_{n \to \infty} \left\{ \lim_{h \to 0} \frac{1 - nh}{1 + nh} \right\} = \lim_{n \to \infty} (1) = 1.$$

General theorems on interchange of limits are beyond the scope of this book. From time to time we shall have to deal with repeated limits of simple functions, and we shall give the most straightforward argument available in each particular case.

6.4. Rate of increase of exp x

We prove the theorem which led us to discuss repeated limits.

Theorem 6.41. $d\,(\exp x)/dx = \exp x.$

Proof. Using theorem 6.2, we have

$$\frac{\exp(x+h) - \exp x}{h} = \exp x \,\frac{\exp h - 1}{h}.$$

Now
$$\frac{\exp h - 1}{h} = 1 + \frac{h}{2!} + \frac{h^2}{3!} + \cdots$$
$$= 1 + \phi(h), \quad \text{say.}$$

We wish to prove that $\phi(h) \to 0$ as $h \to 0$. We have

$$|\phi(h)| \leqslant \frac{|h|}{2} + \frac{|h|^2}{4} + \cdots + \frac{|h|^n}{2^n} + \cdots$$

$$= \frac{|\tfrac{1}{2}h|}{1 - |\tfrac{1}{2}h|} \quad (\text{if } |h| < 2)$$

$$\to 0 \quad \text{as} \quad h \to 0. \ \blacksquare$$

Corollary. exp x *is a continuous function.*

This follows from §4.1 (3). Alternatively it could be proved directly by an argument similar to that used in the theorem.

Theorem 6.42. exp x *is a strictly increasing function and, if* $y = \exp x$, *y takes every value greater than* 0 *for one value of x.*

Proof. If $x \geqslant 0$, then, at once from the series, $\exp x \geqslant 1$.

If $x < 0$, $$\exp x = 1/\exp(-x) > 0.$$

Since $\exp x$ has a derivative which is positive for all values of x, it increases strictly.

As $x \to \infty$, plainly $\exp x \to \infty$.

As $x \to -\infty$, $\exp x = 1/\exp(-x) \to 0+$. $\blacksquare$

Theorem 6.43. (*The order of magnitude of* exp x.)

For any fixed k (*however large*)

$$\frac{\exp x}{x^k} \to \infty \quad as \quad x \to \infty.$$

Proof. Let n be the integer next greater than k.

If $x > 0$, $\exp x > x^n/n!$, since this is just one term of the series defining $\exp x$. $\blacksquare$

You should acquire a vivid appreciation of this important fact. *For large x, the function* exp x *is larger than any power of x* (see fig. 3a, p. 110).

6.5. exp x as a power

Scrutiny of the series does not reveal that $\exp x$ is the xth power of a constant. This fact will be shown to follow from theorem 6.2.

Definition. $e = \exp 1$.

The number e, namely

$$1 + \frac{1}{1!} + \frac{1}{2!} + \ldots + \frac{1}{n!} + \ldots$$

is one of the fundamental constants of mathematics. Its value to ten places of decimals is $2 \cdot 7182818285$. It is easy to prove that e is irrational—the proof is sketched in exercise 6(a), 1. A more difficult argument (outside our range) is required to prove that e is not an algebraic number, that is to say, it is not the root of any algebraic equation with integral coefficients.

Theorem 6.5. *If r is rational, $\exp r = e^r$, where the right-hand side is the positive rth power of the number e.*

Note. To understand why the meaning of e^r is specified, observe that $e^{\frac{1}{2}}$ (say) has two values ($\pm 1 \cdot 6487\ldots$), whereas $\exp \frac{1}{2}$ is uniquely defined (by the series).

Proof. If r is a positive integer n, theorem 6.2 gives

$$\exp n = (\exp 1)^n = e^n.$$

If r is a negative integer $-n$, then

$$\exp(-n) = \frac{1}{\exp n} = \frac{1}{e^n} = e^{-n}.$$

If r is a rational p/q, where p and q are integers, then

$$\{\exp(p/q)\}^q = \exp p \quad \text{(by theorem 6.2)}$$

$$= e^p \quad \text{(just proved)}.$$

Therefore $\exp(p/q) = e^{p/q}$ |.

Irrational powers. What do we mean by (say) $3^{\sqrt{2}}$? It is likely that the reader, if he looks back over the work on indices which has so far been put before him, will find that no meaning has yet been given to it. Numbers like $3^{7/5}$ were defined in such a way as to obey the index laws such as $a^m a^n = a^{m+n}$. These laws do not provide a definition for $3^{\sqrt{2}}$.

It is natural to suggest that, as $\sqrt{2}$ can be approached as closely as we wish by rational numbers (see §1.5), we could define $3^{\sqrt{2}}$ as the limit of 3^r as r runs through a sequence of rationals with $\sqrt{2}$ as limit. This course, though possible, makes heavier going that one might expect. We should need to prove that the numbers 3^r *have* a limit; and, further, that the limit is the same for two different sequences both of which approach $\sqrt{2}$.

We have then to think how best to elucidate the general power a^x. There is one particular value of a for which the preceding work indicates what to do, namely $a = e$.

Definition. *If x is irrational, e^x is defined to mean* $\exp x$.

We have already *proved* that, if x is rational, $e^x = \exp x$. So this equality holds for all values of x.

We postpone until §6.6 the further discussion of a^x when a has a value other than e.

Exercises 6 (a)

Notes on these exercises are given on p. 179.

1. Prove that
$$\sum_{m+1}^{\infty} \frac{1}{n!} < \frac{1}{m(m!)}.$$
Deduce that e is irrational.

2. Prove that, as $n \to \infty$,
$$\left(1 + \frac{1}{n}\right)^n \to e.$$

3. Investigate the limit, as $n \to \infty$, of
$$\left(1 + \frac{x}{n}\right)^n,$$
where x is any real number.

4. Let k be a constant such that $0 < k < 1$. Sketch the graph of the function
$$e^{-x}\left(1 + x + \frac{x^2}{2!} + \ldots + \frac{x^n}{n!}\right) - k.$$

Prove that there is just one positive root x_n of the equation
$$1 + x + \frac{x^2}{2!} + \ldots + \frac{x^n}{n!} = k e^x.$$

Show also that (k remaining fixed) x_n increases as n increases.

5. Find approximately the large root of the equation
$$e^x = x^{1000}.$$

6. Arrange the functions
$$x^{-1}\exp(\sqrt{x}), \quad x\exp\{(\log x)^2\}, \quad x^2\exp\{(\log x)^{\frac{1}{2}}\}$$
in order of magnitude for large values of x.

7. Is it possible to find a function which, as $x \to \infty$, tends to infinity more slowly than $e^{\delta x}$ for every $\delta > 0$ and more rapidly than x^n for every n?

6.6. The logarithmic function

Theorems 3.9 and 6.42 enable us to define a function of a real variable which is the inverse of the exponential function.

Definition. *If $x > 0$, write $y = \log x$ if $x = e^y$.*

The following properties follow from the corresponding properties of the exponential function, and the reader should go through the steps of the deduction. We always suppose that the number following the symbol log is positive.

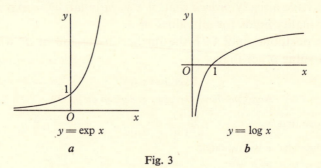

$y = \exp x$

a

$y = \log x$

b

Fig. 3

$\log x$ is continuous and differentiable and

$$\frac{d}{dx}(\log x) = \frac{1}{x}.$$

$$\log(ab) = \log a + \log b,$$

$$\log x \to \infty \quad \text{as} \quad x \to \infty,$$

$$\log x \to -\infty \quad \text{as} \quad x \to 0+.$$

If $k > 0$,

$$\frac{\log x}{x^k} \to 0 \quad \text{as} \quad x \to \infty.$$

This last property (correlative to that at the end of §6.4) is important. In words, $\log x$ tends to infinity as x tends to infinity, but more slowly than x raised to any positive power (however small). The general shape of the graph of $y = \log x$ is shown in fig. 3*b*.

There is a simple representation of $\log(1+x)$ as a power series.

Theorem 6.6. *If* $-1 < x < 1$

$$\log(1+x) = x - \frac{x^2}{2} + \frac{x^3}{3} - \ldots + (-1)^{n-1}\frac{x^n}{n} + \ldots.$$

The proof is left to the reader. Starting from Maclaurin's theorem it follows very closely the proof of the binomial theorem given on p. 102. Lagrange's form of remainder will serve for positive x, but Cauchy's form is to be used for negative x.

Note. The result of theorem 6.6 is also true for $x = 1$. This is most easily proved using integration and following the method of theorem 7.9 (see p. 136).

The general power a^x. At the end of §6.5 we had defined a^x when x is rational and also for irrational x in the special case $a = e$. We have still to attach a meaning to a^x when x is irrational and $a \neq e$. We adopt the following definition.

Definition. *If* $a > 0$ *and* x *is irrational,* a^x *means*

$$e^{x \log a}.$$

This is consistent with the definition of e^x given in §6.5, and the relation $a^x = e^{x \log a}$ holds when x is rational. The reader can verify that the index laws

$$a^x \times a^y = a^{x+y}, \quad (a^x)^y = a^{xy}$$

hold for $a > 0$ whatever the values of x and y.

Exercises 6 (*b*)
Notes on these exercises are given on p. 179.

1. Prove that, as $n \to \infty$,
$$n(x^{1/n} - 1) \to \log x.$$

2. Prove that, if $x > 0$,
$$x - \tfrac{1}{2}x^2 < \log(1+x) < x - \tfrac{1}{2}x^2 + \tfrac{1}{3}x^3.$$

Give an extension of these inequalities.

3. A differentiable function f (not identically 0) satisfies the functional equation
$$f(xy) = f(x) + f(y).$$
Prove that $f'(x) = A/x$.

4. Prove that, if $y > 0$,

$$\log y = 2 \left\{ \frac{y-1}{y+1} + \frac{1}{3} \left(\frac{y-1}{y+1} \right)^3 + \frac{1}{5} \left(\frac{y-1}{y+1} \right)^5 + \ldots \right\}.$$

Use this series to calculate $\log 2$ to three places of decimals.

5. Find the limits, as $x \to 0$ and as $x \to \infty$, of

$$\text{(i)} \quad \frac{\log(1+ax)}{x}, \quad \text{(ii)} \quad \frac{a^x - b^x}{c^x - d^x},$$

where a, b, c, d are positive and $c \neq d$.

6.7. Trigonometric functions

As predicted in §1.2 we base our account of the trigonometric functions on the definitions

$$\cos x = 1 - \frac{x^2}{2!} + \frac{x^4}{4!} - \ldots,$$

$$\sin x = x - \frac{x^3}{3!} + \frac{x^5}{5!} - \ldots.$$

These series are absolutely convergent for all values of x (real or complex). Among the properties that we should expect to establish at an early stage are

$$\sin(x+y) = \sin x \cos y + \cos x \sin y,$$

$$\frac{d}{dx} \sin x = \cos x.$$

The first of these, involving multiplication of series, could be proved by the same principles (though with detail which is a little more troublesome) as theorem 6.2. The proof that the derivative of $\sin x$ is $\cos x$ would follow closely that of theorem 6.41.

Instead of writing out afresh proofs of trigonometric formulae, modelled on those of exponential formulae, it is more satisfying to observe that, in the setting of complex variables, the trigonometric and exponential functions are very closely related.

Writing
$$\exp z = 1 + z + \frac{z^2}{2!} + \ldots + \frac{z^n}{n!} + \ldots,$$

we see that theorem 6.2 is true for complex variables, and, if we define the derivative of $f(z)$ as

$$\lim_{h \to 0} \frac{f(z+h) - f(z)}{h},$$

noting that h can now assume complex values, then theorem 6.41 is also true. The discussion of the exponential function from theorem 6.42 onwards supposed the variable to be real.

6.8. Exponential and trigonometric functions

From the series concerned we see that

$$\exp(iz) = \cos z + i \sin z,$$
$$\exp(-iz) = \cos z - i \sin z,$$

or expressing $\cos z$ and $\sin z$ in terms of the exponential function,

$$\cos z = \tfrac{1}{2}\{\exp(iz) + \exp(-iz)\},$$
$$\sin z = \frac{1}{2i}\{\exp(iz) - \exp(-iz)\}.$$

These formulae, combined with the properties of the exponential function, enable us to develop the results of analytical trigonometry, in so far as they do not involve periodicity or the number π. We append a short list, which will illustrate the procedure

$$\cos(-z) = \cos z, \quad \cos 0 = 1,$$
$$\sin(-z) = -\sin z, \quad \sin 0 = 0.$$

The addition formulae (x and y may be complex)

$$\sin(x+y) = \sin x \cos y + \cos x \sin y,$$
$$\cos(x+y) = \cos x \cos y - \sin x \sin y.$$

To prove, say, the former, we have

$$2i \sin(x+y) = \exp\{i(x+y)\} - \exp\{-i(x+y)\}$$
$$= \exp(ix) \exp(iy) - \exp(-ix) \exp(-iy)$$

(from theorem 6.2)

$$= \tfrac{1}{2}\{\exp(ix) - \exp(-ix)\} \{\exp(iy) + \exp(-iy)\}$$
$$+ \tfrac{1}{2}\{\exp(ix) + \exp(-ix)\} \{\exp(iy) - \exp(-iy)\},$$

8

giving $\qquad \sin(x+y) = \sin x \cos y + \cos x \sin y.$ ▌

$$\cos^2 x + \sin^2 x = 1.$$

From the last result, *if now x is real,*

$$-1 \leqslant \cos x \leqslant 1, \quad -1 \leqslant \sin x \leqslant 1.$$

Periodicity of the trigonometric functions. This is a surprising property. No one could divine from inspection or simple manipulation of the series for $\sin x$ and $\cos x$ that they repeat their values at regular intervals of x. We prove a theorem from which periodicity will follow very easily.

Theorem 6.81. *There is a smallest positive constant* $\frac{1}{2}\varpi$ *(where* $\sqrt{2} < \frac{1}{2}\varpi < \sqrt{3}$) *such that* $\cos\frac{1}{2}\varpi = 0.$

Proof. If $0 < x < 2$,

$$\sin x = \left(x - \frac{x^3}{3!}\right) + \left(\frac{x^5}{5!} - \frac{x^7}{7!}\right) + \dots$$

$$> 0,$$

since the positive term in each bracket is greater than the negative term.

So, for $0 < x < 2$, $\cos x$, having the negative derivative $-\sin x$, is a decreasing function.

The theorem will be proved when we have shown that $\cos\sqrt{2} > 0$ and $\cos\sqrt{3} < 0$. If we bracket the terms of the series for $\cos x$ in pairs, as we did those of $\sin x$, the first bracket is 0 for $x = \sqrt{2}$ and all the others are positive.

Again,

$$\cos x = 1 - \frac{x^2}{2!} + \frac{x^4}{4!} - \left(\frac{x^6}{6!} - \frac{x^8}{8!}\right) - \dots,$$

the succeeding terms being bracketed in pairs.

When

$$x = \sqrt{3}, \quad 1 - \frac{x^2}{2!} + \frac{x^4}{4!} = 1 - \tfrac{3}{2} + \tfrac{3}{8} < 0$$

and, as before, the first term in each bracket exceeds the second. ▌

Corollary. $\sin\frac{1}{2}\varpi = 1.$

For $\sin^2\frac{1}{2}\varpi + \cos^2\frac{1}{2}\varpi = 1$ and $\sin\frac{1}{2}\varpi$ is positive.

Theorem 6.82. *If ϖ is the number defined in the preceding theorem, then, for all values of x,*

(1) $\sin(x + \frac{1}{2}\varpi) = \cos x$, $\cos(x + \frac{1}{2}\varpi) = -\sin x$;

(2) $\sin(x + \varpi) = -\sin x$, $\cos(x + \varpi) = -\cos x$;

(3) $\sin(x + 2\varpi) = \sin x$, $\cos(x + 2\varpi) = \cos x$.

The proof is immediate from the addition theorems. We have thus shown that the functions $\sin x$ and $\cos x$ have period 2ϖ, and it is easy to see that no smaller number is a period.

In these two theorems we have adopted the notation ϖ (an alternative way of writing the Greek letter *pi*). The number ϖ will (in §7.9) be identified with the ratio of the circumference of a circle to its diameter. Anticipating this identification, we shall henceforward write π instead of ϖ.

The remaining trigonometric (or circular) functions are defined in terms of the sine and cosine in the usual way

$$\tan z = \frac{\sin z}{\cos z}, \quad \cot z = \frac{1}{\tan z}, \quad \sec z = \frac{1}{\cos z}, \quad \operatorname{cosec} z = \frac{1}{\sin z}.$$

From the periodicity of the sine and cosine, the relation

$$\exp(iz) = \cos z + i \sin z$$

shows at once that $\exp z$ has the (imaginary) period $2\pi i$.

Exercises 6 (c)

Notes on these exercises are given on pp. 179–80.

1. Refine the argument of theorem 6.81 to give closer bounds for $\frac{1}{2}\pi$, for example, $1 \cdot 5 < \frac{1}{2}\pi < 1 \cdot 6$. (The series for $\cos x$ and $\sin x$ do not provide a practical way of obtaining accurate approximations to π. For better methods see §7.9.)

2. Prove the statement following theorem 6.82 that 2π is the smallest period of $\cos x$ and $\sin x$.

3. If a and b are positive constants and x is real, prove that

$$f(x) = a \cot x + b \operatorname{cosec} x$$

takes all real values if $a > b$, and all real values except for a certain range if $a < b$.

Sketch graphs of $y = f(x)$ for $a > b$, $a = b$, $a < b$.

4. Prove that, as $n \to \infty$,

$$\left(\cos\frac{\alpha}{n}\right)^{n^2} \to e^{-\frac{1}{2}\alpha^2}.$$

6.9. The inverse trigonometric functions

Theorem 6.91. *The equation* $x = \sin y$ *defines an inverse function written*

$$y = \text{arc } \sin x$$

such that y increases from $-\frac{1}{2}\pi$ *to* $\frac{1}{2}\pi$ *as x increases from* -1 *to* 1. *Also*

$$\frac{dy}{dx} = \frac{1}{\sqrt{(1-x^2)}}.$$

Proof. $dx/dy = \cos y > 0$ for $-\frac{1}{2}\pi < y < \frac{1}{2}\pi$. Therefore $\sin y$ strictly increases from -1 to 1 as y increases from $-\frac{1}{2}\pi$ to $\frac{1}{2}\pi$. By theorem 3.9, there is an inverse function with the range of values stated. Also, by §4.2 (6),

$$\frac{dy}{dx} = \frac{1}{dx/dy} = \frac{1}{\cos y} = \frac{1}{\sqrt{(1-x^2)}},$$

where the positive square root is to be taken since $\cos y$ is positive for y between $-\frac{1}{2}\pi$ and $\frac{1}{2}\pi$. |

Notes on theorem 6.91. (1) The equation $x = \sin y$ defines infinitely many values of y for each x such that $-1 \leqslant x \leqslant 1$. For, by the periodicity of $\sin y$, any integral multiple of 2π can be added to the value of y which lies between $-\frac{1}{2}\pi$ and $\frac{1}{2}\pi$. In the theorem we have singled out the *principal value* of arc sin x.

(2) Similarly the equation $x = \cos y$ defines $y = \text{arc cos } x$, where y decreases from π to 0 as x increases from -1 to 1. Here

$$\frac{dy}{dx} = -\frac{1}{\sqrt{(1-x^2)}}.$$

Theorem 6.92. *The equation* $x = \tan y$ *defines an inverse function* $y = \text{arc tan } x$ *defined for all x. y is an increasing function and*

$$\lim_{x \to -\infty} y = -\tfrac{1}{2}\pi, \quad \lim_{x \to \infty} y = \tfrac{1}{2}\pi.$$

Also

$$\frac{dy}{dx} = \frac{1}{1+x^2}.$$

Proof. $dx/dy = \sec^2 y = 1 + x^2$.

$\tan y$ is a continuous increasing function for $-\frac{1}{2}\pi < y < \frac{1}{2}\pi$ and, as $y \to \pm\frac{1}{2}\pi$, $x \to \pm\infty$, respectively. The existence of the inverse function and the value of its derivative follow from theorem 3.9 and §4.2 (6). |

Note. As in theorem 6.91, we are defining a principal value. If, for a given value x_1 of x, the value $y = y_1$ satisfies $x = \tan y$, so does $y_1 + n\pi$, where n is any positive or negative integer.

6.10. The hyperbolic functions and their inverses

Define the hyperbolic cosine and sine by the formulae

$$\cosh z = \tfrac{1}{2}(e^z + e^{-z}),$$
$$\sinh z = \tfrac{1}{2}(e^z - e^{-z}).$$

Thus $\quad \cosh z = \cos iz \quad$ and $\quad i \sinh z = \sin iz.$

In nearly all applications z is real. The functions get their names from the fact that the point $x = \cosh t$, $y = \sinh t$ traces out a branch of the hyperbola $x^2 - y^2 = 1$ (whereas $x = \cos t$, $y = \sin t$ traces out the circle $x^2 + y^2 = 1$).

The hyperbolic functions have properties analogous to those of the circular functions cosine and sine. The following are among the most useful.

$$\frac{d}{dx}\cosh x = \sinh x, \quad \frac{d}{dx}\sinh x = \cosh x,$$

$$\cosh^2 x - \sinh^2 x = 1,$$

$$\cosh^2 x + \sinh^2 x = \cosh 2x,$$

$$2\cosh x \sinh x = \sinh 2x.$$

The reader can supply proofs and can construct other formulae. He should also sketch graphs of the functions.

The inverse functions will be useful in integration. If $x = \cosh y$, then from the definition

$$e^{2y} - 2xe^y + 1 = 0,$$

giving $\qquad e^y = x \pm \sqrt{(x^2 - 1)}$

and so $\qquad y = \log\{x \pm \sqrt{(x^2 - 1)}\},$

which is the same as

$$y = \pm\log\{x + \sqrt{(x^2 - 1)}\}.$$

If we take the $+$ sign, y is defined for all $x \geqslant 1$ and is denoted by $y = \text{arg}\cosh x$.

Then
$$\frac{dy}{dx} = \frac{1}{dx/dy} = \frac{1}{\sinh y} = \frac{1}{\sqrt{(x^2-1)}}.$$

For $\sinh x$ the discussion is simpler. If $x = \sinh y$, then x increases from $-\infty$ to ∞ as y increases from $-\infty$ to ∞ and there is a unique inverse function

$$y = \arg \sinh x.$$

or, as a logarithm,

$$y = \log \{x + \sqrt{(x^2+1)}\}.$$

Exercises 6 (d)

Notes on these exercises are given on p. 180.

1. Find the sum $\sum\limits_{r=0}^{n} \cosh(\alpha + 2r\beta)$.

2. Prove that the equation
$$x = 2 + \log x$$

has two positive roots, say a and b.
  The sequence x_n is defined by
$$x_{n+1} = 2 + \log x_n \quad (n = 1, 2, 3, ...),$$

where $a < x_1 < b$. Prove that $x_n \to b$.

3. Prove that, if $-1 < r < 1$,
$$1 + r \cos \theta + ... + r^n \cos n\theta + ... = \frac{1 - r \cos \theta}{1 - 2r \cos \theta + r^2},$$

$$r \sin \theta + ... + r^n \sin n\theta + ... = \frac{r \sin \theta}{1 - 2r \cos \theta + r^2}.$$

4. Investigate results like those of **3** for hyperbolic functions.

5. Prove that, if $y = e^{-\frac{1}{2}x^2}$, then
$$y^{(n+2)} + xy^{(n+1)} + (n+1)y^{(n)} = 0.$$

The functions f_n are defined by
$$f_n(x) = (-1)^n e^{\frac{1}{2}x^2} \frac{d^n}{dx^n} (e^{-\frac{1}{2}x^2}).$$

Prove that
  (i) $f'_{n+1} = (n+1)f_n$,
  (ii) $f_{n+1} = xf_n - f'_n$,
  (iii) $f_{n+2} - xf_{n+1} + (n+1)f_n = 0$,
  (iv) f_n is a polynomial of degree n.

6. Obtain expansions in power-series of the following functions. Find the general term if you can, otherwise the first three non-vanishing terms.
  (i) $\cos^3 x$,   (ii) $\tan x$,
  (iii) $(\arcsin x)^2$,   (iv) $\sin (m \arcsin x)$,
  (v) $e^x \cos 2x$,   (vi) $\cos \log (1+x)$.

7

THE INTEGRAL CALCULUS

7.1. Area and the integral

Historically the concept of a definite integral was developed to represent an area bounded by curved lines. This geometrical equivalence helps one to visualise the meaning of the analytical expressions which occur in the definition and manipulation of integrals.

Let f be a function defined in (a, b). The area which we proceed to measure is that bounded by the curve $y = f(x)$, the ordinates $x = a$ and $x = b$, and the x-axis. (There may be a gain in clarity if you think of f as taking only positive values in (a, b), but the following analysis holds if f can take values of either sign.) All that we assume for the present about f is that it is bounded. In practice f is usually continuous, and that assumption will be made at the stage at which it simplifies the discussion.

Our method will be to obtain approximations from above and from below to the curved area which we wish to measure. We must start with a number of definitions.

Definitions. Given an interval (a, b), then a finite set of numbers $a, x_1, x_2, ..., x_{n-1}, b$, such that

$$a < x_1 < x_2 < ... < x_{n-1} < b$$

is called a *dissection* of (a, b). Each x_r is a *point of division*.

To complete the scheme of suffixes we can write $a = x_0$ and $b = x_n$.

Each of the intervals (x_{r-1}, x_r) for $r = 1, 2, ..., n$ is a *subinterval* of the dissection. Let δ_r be the length of the rth subinterval,

$$\delta_r = x_r - x_{r-1}.$$

The length of the greatest subinterval

$$\delta^* = \max \delta_r$$

is called the *norm* of the dissection.

We next define upper and lower *approximative sums*.

Suppose that M_r and m_r are respectively the supremum and infimum of $f(x)$ for x in the rth subinterval taken closed, i.e. for

$$x_{r-1} \leqslant x \leqslant x_r.$$

Write

$$S = \sum_{r=1}^{n} M_r \delta_r$$

and

$$s = \sum_{r=1}^{n} m_r \delta_r.$$

Then the upper sum S is the sum of the areas of n rectangles, of which the rth has base (x_{r-1}, x_r) and height M_r. The sum of these areas is greater than or equal to the area R contained between the curve $y = f(x)$ and the lines $x = a$, $x = b$, $y = 0$. Similarly the lower sum s is less than or equal to the area R.

If now ξ_r is any value of x in the rth subinterval,

$$x_{r-1} \leqslant \xi_r \leqslant x_r \quad (r = 1, 2, ..., n)$$

and if we form the sum

$$\sigma = \sum_{r=1}^{n} f(\xi_r)\, \delta_r,$$

then

$$s \leqslant \sigma \leqslant S.$$

Our ultimate aim is to prove that, if f is a function of one of the commonly occurring types (including continuous functions and monotonic functions), the sum σ tends to a limit as δ^* tends to 0. We shall define this limit to be the integral of the function f over (a, b).

You will observe that this limiting process is of a less simple kind than those which we have so far encountered. If the function f is given, the number σ depends on the x_r and the ξ_r; in fact, σ is a function whose domain is a set of dissections. In the passage to the limit, we suppose δ^* to take a sequence of values tending to 0, and the permitted dissections are progressively restricted by the requirement that their norms are to be less than δ^*.

On account of the complexity of the limit-operation, we first take an easier approach to the integral through the *bounds* of the sums S, s for all dissections.

7.2. The upper and lower integrals

If M, m are the supremum and infimum of $f(x)$ in $a \leqslant x \leqslant b$, and if, given any dissection $\mathscr{D}$, we construct the sums S, s as above, calling them $S(\mathscr{D})$ and $s(\mathscr{D})$, it follows from the inequalities

$$M_r \leqslant M \quad \text{and} \quad m_r \geqslant m \quad (r = 1, 2, \ldots, n)$$

that $\qquad m(b-a) \leqslant s(\mathscr{D}) \leqslant S(\mathscr{D}) \leqslant M(b-a)$.

So the set of numbers $S(\mathscr{D})$ corresponding to all dissections $\mathscr{D}$ of (a, b), being bounded below by $m(b-a)$, has an infimum, J say. Similarly the set of numbers $s(\mathscr{D})$ has a supremum j.

Our aim, achieved by theorem 7.22, is to prove that $J \geqslant j$.

Theorem 7.21. *The introduction of a new point of division decreases the upper sum S.*

Proof. Suppose that $S(\mathscr{D}_1)$ is the upper sum for the dissection $\mathscr{D}_1$. Let the dissection $\mathscr{D}_2$ be formed from $\mathscr{D}_1$ by the introduction of a new point of division x_r' into the interval (x_{r-1}, x_r). Let M_r', M_r'' be the suprema of $f(x)$ in the closed intervals (x_{r-1}, x_r'), (x_r', x_r), respectively. Then $M_r' \leqslant M_r$ and $M_r'' \leqslant M_r$.

The contribution of the interval (x_{r-1}, x_r) to $S(\mathscr{D}_1)$ is $M_r(x_r - x_{r-1})$. Its contribution to $S(\mathscr{D}_2)$ is

$$M_r'(x_r' - x_{r-1}) + M_r''(x_r - x_r')$$
$$\leqslant M_r(x_r - x_{r-1}).$$

Since the contribution to $S(\mathscr{D}_1)$ and $S(\mathscr{D}_2)$ of each subinterval except (x_{r-1}, x_r) is the same, we have

$$S(\mathscr{D}_2) \leqslant S(\mathscr{D}_1).$$

Corollary. If $\mathscr{D}_1$, $\mathscr{D}_2$ are dissections of (a, b) for which every point of $\mathscr{D}_1$ is a point of $\mathscr{D}_2$, then

$$S(\mathscr{D}_2) \leqslant S(\mathscr{D}_1)$$

and similarly $\qquad s(\mathscr{D}_2) \geqslant s(\mathscr{D}_1)$.

Definition. If $\mathscr{D}_1$ and $\mathscr{D}_2$ are related as in the corollary, we may call $\mathscr{D}_2$ a *refinement* of $\mathscr{D}_1$.

Theorem 7.22. $J \geqslant j$.

Proof. Let $\mathscr{D}_1$ and $\mathscr{D}_2$ be any two dissections of (a, b).

Let $\mathscr{D}_3$ be the dissection whose points of division are all the points of division of either $\mathscr{D}_1$ or $\mathscr{D}_2$. So $\mathscr{D}_3$ is a refinement both of $\mathscr{D}_1$ and of $\mathscr{D}_2$. The corollary to the last theorem gives

$$S(\mathscr{D}_3) \leqslant S(\mathscr{D}_1) \quad \text{and} \quad s(\mathscr{D}_3) \geqslant s(\mathscr{D}_2).$$

But
$$S(\mathscr{D}_3) \geqslant s(\mathscr{D}_3),$$

being upper and lower sums for the same dissection. Combining all these inequalities we have

$$S(\mathscr{D}_1) \geqslant s(\mathscr{D}_2).$$

Since this is true for all dissections $\mathscr{D}_1$,

$$J = \inf S(\mathscr{D}_1) \geqslant s(\mathscr{D}_2).$$

Since the last line is true for all dissections $\mathscr{D}_2$,

$$J \geqslant \sup s(\mathscr{D}_2) = j. \quad \blacksquare$$

If we assume about f only what we have assumed already, that it is a bounded function, then it is possible for J to be greater than j or equal to j.

Illustration. An example in which $J > j$. Define

$$f(x) = 1 \quad \text{if} \quad x \text{ is rational,}$$
$$f(x) = 0 \quad \text{if} \quad x \text{ is irrational,}$$

and take the interval (a, b) to be $(0, 1)$.

Then, whatever dissection $\mathscr{D}$ is taken, every $M_r = 1$ and $S(\mathscr{D}) = 1$. Every $m_r = 0$ and $s(\mathscr{D}) = 0$. So $J = 1$ and $j = 0$.

Here $y = f(x)$, far from representing a curve in a straightforward sense capable of bounding an area, is discontinuous for every value of x.

Exercise. Give the simplest example you can think of in which $J = j$.

The numbers J, j, being approximations from above and from below to our intuitive notion of an integral, are often called upper and lower integrals. They can be represented by the usual integral sign with a bar above or below. We shall not go into further detail about upper and lower integrals, but shall confine ourselves to the most useful case in which $J = j$ and there is an integral in the ordinary sense.

7.3. The integral as a limit

Definition. If, with the notation of §7.1,

$$\sigma = \sum_{r=1}^{n} f(\xi_r)\,\delta_r$$

tends to a limit as $\delta^* \to 0$, then f is said to be *integrable* in (a, b) and the limit is written

$$\int_a^b f(x)\,dx \quad \text{or} \quad \int_a^b f.$$

The latter, shorter form is usually appropriate in the discussion of general properties (e.g. those of §7.5). If a particular function is being integrated, the specification of it must take the form

$$\int_a^b f(x)\,dx, \quad \text{e.g.} \int_0^1 (x^2 + 3)\,dx.$$

Theorem 7.31. *If $S - s \to 0$ as $\delta^* \to 0$, then f is integrable in (a, b).*

Proof. Given ϵ, there is δ such that

$$S - s < \epsilon$$

for any dissection such that $\delta^* < \delta$.

Now

$$S - s = (S - J) + (J - j) + (j - s),$$

and each term on the right-hand side is greater than or equal to 0.

Then

$$J - j \leqslant S - s < \epsilon$$

if the norm of the dissection is less than δ.

But J and j do not depend on ϵ, and therefore

$$J - j = 0.$$

Thus both S and s tend to J as $\delta^* \to 0$, and so does σ, which lies between S and s. ∎

The converse of the theorem also holds.

Theorem 7.32. *If f is integrable in (a, b), then*

$$S - s \to 0 \quad \text{as} \quad \delta^* \to 0.$$

Proof. If I is the value of the integral, then, given ϵ, there is δ such that, if $\mathscr{D}$ is any dissection of norm less than δ,

$$I-\epsilon < \sum_{r=1}^{n} f(\xi_r) \, \delta_r < I+\epsilon,$$

where ξ_r is an arbitrary point of δ_r.

If M_r is the supremum of f in δ_r, we can choose ξ_r such that

$$f(\xi_r) > M_r - \frac{\epsilon}{n}.$$

Therefore

$$\sum_{r=1}^{n} M_r \delta_r < \sum_{r=1}^{n} f(\xi_r) \, \delta_r + \epsilon$$

$$< I + 2\epsilon.$$

Similarly

$$\sum_{r=1}^{n} m_r \delta_r > I - 2\epsilon.$$

Hence $S - s < 4\epsilon$, and so $S - s \to 0$ as $\delta^* \to 0$. |

7.4. Continuous or monotonic functions are integrable

The condition of theorem 7.31 is easy to establish for continuous functions and for monotonic functions.

Theorem 7.41. *A function f continuous in the closed interval (a, b) is integrable.*

Proof. By theorem 3.82, given ϵ, there is δ such that

$$M_r - m_r < \frac{\epsilon}{b-a}$$

for every subinterval of any dissection $\mathscr{D}$ with norm less than δ. Then

$$S(\mathscr{D}) - s(\mathscr{D}) = \Sigma(M_r - m_r) \, \delta_r$$

$$< \frac{\epsilon}{b-a} \Sigma \delta_r = \epsilon. \ |$$

Theorem 7.42. *A function f monotonic in the closed interval (a, b) is integrable.*

Proof. We may suppose f increasing.

Then, in $x_{r-1} \leqslant x \leqslant x_r$,

$$M_r = f(x_r) \quad \text{and} \quad m_r = f(x_{r-1}).$$

So
$$S - s = \Sigma(M_r - m_r)\,\delta_r$$
$$\leqslant \delta^* \Sigma(M_r - m_r)$$
$$= \delta^*\{f(b) - f(a)\}. \quad \blacksquare$$

Exercises 7 (a)

Notes on these exercises are given on p. 180.

Calculation of simple integrals from the definition.

1. Calculate $\int_0^1 x\,dx$ by dissecting $(0, 1)$ into n equal parts.

2. Calculate $\int_a^b x^k\,dx$, where $k > 0$, by dividing (a, b) into n parts in geometric progression at the points $aq, aq^2, \ldots, aq^{n-1}$, where $aq^n = b$.

3. Calculate $\int_0^\alpha \sin x\,dx$ by dissecting $(0, \alpha)$ into equal parts.

4. By the method of 2, prove that
$$\int_1^2 \frac{dx}{x^2} = \frac{1}{2}.$$

Deduce that
$$\lim_{n\to\infty} n\left\{\frac{1}{(n+1)^2} + \frac{1}{(n+2)^2} + \ldots + \frac{1}{(2n)^2}\right\} = \frac{1}{2}.$$

A theorem of Darboux.

5. With the notation of §7.1, $S \to J$ as $\delta^* \to 0$.

This is an important theorem, showing that the number J which was defined as a bound is in fact a limit. As the proof is a little more difficult than those of any results in the text, we have developed properties of the integral independently of it.

7.5. Properties of the integral

In defining the integral we supposed that $a < b$. If $a > b$, we define
$$\int_a^b f(x)\,dx = -\int_b^a f(x)\,dx.$$

The following properties are constantly used.

(1) *If* $a \leqslant c < d \leqslant b$ *and* f *is integrable in* (a, b), *then* f *is integrable in* (c, d).

Proof. Given ϵ, there is from theorem 7.32 δ such that, if $\mathscr{D}$ is any dissection of (a, b) with norm less than δ, then
$$S(\mathscr{D}) - s(\mathscr{D}) < \epsilon.$$

Let $\mathscr{D}'$ be any dissection of (c, d) with norm less than δ. By adding appropriate points of division in (a, c) and (d, b), we obtain a dissection $\mathscr{D}$ of (a, b), of which $\mathscr{D}'$ is a 'part'.

$$S(\mathscr{D}') - s(\mathscr{D}') = \Sigma(M_r - m_r)\,\delta_r$$

summed over the subintervals of $\mathscr{D}'$.

All the terms on the right-hand side are contained in $S(\mathscr{D}) - s(\mathscr{D})$ and so

$$S(\mathscr{D}') - s(\mathscr{D}') \leqslant S(\mathscr{D}) - s(\mathscr{D}) < \epsilon$$

provided only that the norm of $\mathscr{D}'$ is less than δ. Therefore, by theorem 7.31, $\int_c^d f$ exists.

(2) *If $a < c < b$, and f is integrable in (a, b), then*

$$\int_a^b f = \int_a^c f + \int_c^b f.$$

Proof. Let $\mathscr{D}$ be a dissection of (a, b) having c as one of its points of division.

Then, in a notation which explains itself,

$$\sum_{(a,b)} f(\xi_r)\,\delta_r = \sum_{(a,c)} f(\xi_r)\,\delta_r + \sum_{(c,b)} f(\xi_r)\,\delta_r.$$

Take the limit as the norm of $\mathscr{D}$ tends to 0. ∎

(3) *If k is a constant,*

$$\int_a^b kf = k\int_a^b f.$$

The proof is easy.

(4) *If f and g are integrable in (a, b), so is their sum*

$$s = f + g,$$

and
$$\int_a^b s = \int_a^b f + \int_a^b g.$$

Proof. For any dissection $\mathscr{D}$ and any choice of ξ_r in δ_r

$$\sum_{r=1}^n s(\xi_r)\,\delta_r = \sum_{r=1}^n f(\xi_r)\,\delta_r + \sum_{r=1}^n g(\xi_r)\,\delta_r.$$

Each of the two sums on the right-hand side tends to the

corresponding integral as $\delta^* \to 0$. Therefore $\int_a^b s$ exists and is equal to the sum of the integrals of f and g. $\blacksquare$

In inequalities such as (5) and in other contexts in which the sense requires it, we suppose $a < b$. In a statement such as (5) it is plain that f is understood to be integrable and we need not say so explicitly.

(5) *If $m \leqslant f \leqslant M$, then*

$$m(b-a) \leqslant \int_a^b f \leqslant M(b-a).$$

Proof. For every dissection

$$m(b-a) \leqslant \Sigma f(\xi_r)\,\delta_r \leqslant M(b-a). \quad \blacksquare$$

Corollary 1. If $f \geqslant 0$, then

$$\int_a^b f \geqslant 0.$$

Corollary 2. If f is continuous, then, for some ξ in (a, b),

$$\int_a^b f = f(\xi)\,(b-a).$$

(6) *If f and g are integrable in (a, b), so is the product fg.*

Proof. If we prove that the square of an integrable function is integrable, the result will follow from (3), (4) and the identity

$$4fg = (f+g)^2 - (f-g)^2.$$

We will prove, then, that f^2 is integrable in (a, b).

Let M_r and m_r be the sup and inf of f in δ_r. Given ϵ, there is δ^* such that

$$\Sigma(M_r - m_r)\,\delta_r < \epsilon \quad \text{if} \quad \max \delta_r < \delta^*.$$

If $K = \sup |f|$ in (a, b), we have

$$M_r^2 - m_r^2 \leqslant 2K(M_r - m_r).$$

Hence $\quad \Sigma(M_r^2 - m_r^2)\,\delta_r < 2K\epsilon \quad \text{if} \quad \max \delta_r < \delta^*,$

and this implies the integrability of f^2. $\blacksquare$

(7) (Extension of (5)). *If, further, $g \geqslant 0$, then*

$$m\int_a^b g \leqslant \int_a^b fg \leqslant M\int_a^b g.$$

Corollary. If f is continuous, then

$$\int_a^b fg = f(\xi) \int_a^b g.$$

Proof. Using corollary 1 of (5), we have

$$\int_a^b \{f-m\} g \geqslant 0.$$

(8) *Schwarz's inequality*

$$\left(\int_a^b fg \right)^2 \leqslant \left(\int_a^b f^2 \right) \left(\int_a^b g^2 \right).$$

Proof. This may be deduced from Cauchy's inequality (exercise 1(*d*), 6) or from the fact that

$$\int_a^b (\lambda f + \mu g)^2 = \lambda^2 \int_a^b f^2 + 2\lambda\mu \int_a^b fg + \mu^2 \int_a^b g^2$$

is greater than or equal to 0 for all values of the constants λ, μ.

7.6. Integration as the inverse of differentiation

Suppose that f is integrable in (a, b) and write

$$F(x) = \int_a^x f(t) \, dt \quad (a \leqslant x \leqslant b).$$

Theorem 7.61. *F is a continuous function.*
Proof.
$$F(x+h) - F(x) = \int_x^{x+h} f(t) \, dt.$$

With the notation of §7.5 (5), the absolute value of the right-hand side does not exceed max $(|mh|, |Mh|)$. So

$$F(x+h) - F(x) \to 0 \text{ as } h \to 0. \quad \blacksquare$$

If we make the additional assumption of continuity of f, we can prove a sharper result.

Theorem 7.62. *If f is integrable in (a, b), then, for any value of x for which f is continuous,*
$$F'(x) = f(x).$$

Proof. Suppose that $h > 0$. Let now

$$M = \sup f(t), \quad m = \inf f(t)$$

for $x \leqslant t \leqslant x+h$. M and m then depend on x and h. By §7.5 (5),

$$mh \leqslant \int_x^{x+h} f(t) \, dt \leqslant Mh.$$

Therefore $$m \leqslant \frac{F(x+h)-F(x)}{h} \leqslant M.$$

Let $h \to 0+$. Since f is continuous at x, both m and M tend to $f(x)$. A similar argument holds when $h \to 0-$. ▌

Theorem 7.63. *Let f be continuous in (a, b). Suppose that ϕ is a function having the property*

$$\phi'(x) = f(x) \quad for \quad a \leqslant x \leqslant b.$$

Then $$\int_a^x f(t) \, dt = \phi(x) - \phi(a) \quad for \quad a \leqslant x \leqslant b.$$

Proof. From theorem 7.62, the function $F - \phi$ has derivative 0 for $a \leqslant x \leqslant b$.

By theorem 4.61 (corollary 1), $F - \phi$ is a constant and, since $F(a) = 0$, we have $$F(x) = \phi(x) - \phi(a). ▌$$

The *existence* of a function ϕ having a given continuous derivative f is established by theorem 7.62. Such a function which, by theorem 7.63, is determined except for an additive constant is called *an indefinite integral* of f, written

$$\int f(x) \, dx.$$

An indefinite integral may or may not be readily expressible in terms of known functions; if it is, theorem 7.63 provides the normal method of calculating the *definite integral*

$$\int_a^b f(x) \, dx.$$

A number of illustrative examples follow in §§7.7, 7.8.

7.7. Integration by parts and by substitution

The systematic search for a function if we are given its derivative employs methods which you are likely to know from

9

your earlier work in the calculus. The emphasis is on technique and not on foundations of analysis, and we shall treat it rather summarily. In this section we shall state two general methods of which repeated use will be made. They both arise from the 'inversion' of a formula in differentiation to yield a formula in integration.

Integration by parts. This is the inverse operation of differentiating a product.

If u, v are functions of x, then

$$\frac{d(uv)}{dx} = u\frac{dv}{dx} + v\frac{du}{dx}.$$

By integrating, we have

If du/dx, dv/dx are continuous, then

$$\int u\frac{dv}{dx}\,dx = uv - \int v\frac{du}{dx}\,dx.$$

Integration by substitution. This comes from the result of differentiating a function of a function

$$\frac{dy}{dt} = \frac{dy}{dx}\frac{dx}{dt}.$$

Writing $dy/dx = f(x)$ and $x = g(t)$, we have

If $f(x)$ and $g'(t)$ are continuous then

$$\int f(x)\,dx = \int f\{g(t)\}\,g'(t)\,dt.$$

Taylor's theorem with remainder an integral. By integrating an appropriate integral by parts we can establish another form of the nth order mean value theorem (§4.8) which is sometimes useful. For brevity we replace the a of theorems 4.81 and 4.82 by 0.

Theorem 7.7. *Let $f^{(n)}$ be continuous for $0 \leqslant x \leqslant h$. Then*

$$f(h) = f(0) + \dots + \frac{h^{n-1}}{(n-1)!}f^{(n-1)}(0) + R_n,$$

where $$R_n = \frac{h^n}{(n-1)!}\int_0^1 (1-t)^{n-1} f^{(n)}(th)\,dt.$$

Proof. By substituting $th = u$,

$$R_n = \frac{1}{(n-1)!}\int_0^h (h-u)^{n-1} f^{(n)}(u)\,du.$$

Integrate by parts and we have

$$R_n = -\frac{h^{n-1}}{(n-1)!} f^{(n-1)}(0) + \frac{1}{(n-2)!} \int_0^h (h-u)^{n-2} f^{(n-1)}(u) \, du.$$

The last integral is R_{n-1} in our notation. If we integrate $n-1$ times by parts we arrive at

$$R_n = -\frac{h^{n-1}}{(n-1)!} f^{(n-1)}(0) - \ldots - hf'(0) + \int_0^h f'(u) \, du.$$

Write $f(h) - f(0)$ for the last integral and rearrange the terms. ∎

Applying §7.5 (5) corollary 2 in two different ways to theorem 7.7, we first reconstruct the term in h^n given in theorem 4.82 (assuming however the continuity and not merely the existence of $f^{(n)}$). The second corollary gives the remainder term of theorem 4.83.

Corollary 1.

$$R_n = \frac{h^n}{(n-1)!} f^{(n)}(\theta h) \int_0^1 (1-t)^{n-1} \, dt = \frac{h^n}{n!} f^{(n)}(\theta h).$$

Corollary 2.

$$R_n = \frac{h^n}{(n-1)!} (1-\theta)^{n-1} f^{(n)}(\theta h).$$

7.8. The technique of integration

We recapitulate the methods in most common use, illustrated by examples.

Rational functions. To integrate a rational function, put it into partial fractions. A real root a of the denominator gives constant multiples of $(x-a)^{-n}$, where $n \geqslant 1$.

If $n > 1$,

$$\int \frac{dx}{(x-a)^n} = \frac{(x-a)^{1-n}}{1-n}.$$

If $n = 1$,

$$\int \frac{dx}{x-a} = \begin{cases} \log(x-a) & (x > a), \\ \log(a-x) & (x < a), \end{cases}$$

or, conveniently, in one formula,

$$\int \frac{dx}{x-a} = \log|x-a|.$$

A fraction whose denominator is a quadratic with complex roots is integrated as follows.

Example.

$$\int \frac{4x-1}{3x^2-4x+5}\, dx.$$

The derivative of the denominator is $6x-4$. Write the numerator $4x-1$ as $\frac{2}{3}(6x-4)+\frac{5}{3}$. The first term gives

$$\frac{2}{3}\int \frac{6x-4}{3x^2-4x+5}\, dx = \frac{2}{3}\log (3x^2-4x+5).$$

The second gives

$$\frac{5}{9}\int \frac{dx}{x^2-\frac{4}{3}x+\frac{5}{3}} = \frac{5}{9}\int \frac{dx}{(x-\frac{2}{3})^2+\frac{11}{9}}$$

$$= \frac{5}{9}\frac{3}{\sqrt{11}}\arctan \frac{x-\frac{2}{3}}{\frac{1}{3}\sqrt{11}} = \frac{5}{3\sqrt{11}}\arctan \frac{3x-2}{\sqrt{11}}.$$

To integrate a partial fraction of the form

$$\frac{px+q}{(x^2+2ax+b)^n} \quad (n > 1),$$

write the numerator as $p(x+a)+(q-ap)$ and the problem is reduced to integrating $1/(x^2+2ax+b)^n$. This is achieved by successive 'reduction' of the index n; see the paragraph below on reduction formulae.

Trigonometric functions. (*a*) To integrate a product of cosines and sines, turn the products into sums and differences by using formulae like

$$2\cos a \cos b = \cos(a+b)+\cos(a-b).$$

Examples. (i) $\int \cos^2 x \cos 4x\, dx$.

(ii) $\int \sin^2 x \cos^3 x\, dx$.

In (i), $\cos^2 x \cos 4x = \frac{1}{2}(1+\cos 2x)\cos 4x$

$$= \frac{1}{2}\cos 4x+\frac{1}{4}\cos 6x+\frac{1}{4}\cos 2x.$$

Hence the integral is $\frac{1}{8}\sin 4x+\frac{1}{24}\sin 6x+\frac{1}{8}\sin 2x$.

In (ii), the odd power of $\cos x$ suggests the substitution $u = \sin x$, giving $\int u^2(1-u^2)\, du$ and so $\frac{1}{3}\sin^3 x-\frac{1}{5}\sin^5 x$.

(*b*) The integral of any rational function of $\cos x$ and $\sin x$ can be transformed into the integral of a rational algebraic function by the substitution $\tan\frac{1}{2}x = t$ for which

$$\cos x = \frac{1-t^2}{1+t^2}, \quad \sin x = \frac{2t}{1+t^2}, \quad \frac{dx}{dt} = \frac{2}{1+t^2}.$$

Example.

$$\int \frac{dx}{5+4\cos x} = \int \frac{2dt}{9+t^2} = \tfrac{2}{3} \text{ arc tan } (\tfrac{1}{3} \tan \tfrac{1}{2}x).$$

Reduction formulae. Suppose that we require

$$\int \sin^n x \, dx.$$

This comes under (*a*) above, but it is more easily found by step-by-step reduction of the index *n*. A reduction formula is nearly always derived by the appropriate integration by parts. Here

$$\int \sin^n x \, dx = \int (\sin^{n-1} x) \sin x \, dx$$

$$= -\sin^{n-1} x \cos x + \int (n-1) \sin^{n-2} x \cos x \cos x \, dx$$

$$= -\sin^{n-1} x \cos x + (n-1) \int \sin^{n-2} x (1 - \sin^2 x) \, dx,$$

and so

$$n \int \sin^n x \, dx = -\sin^{n-1} x \cos x + (n-1) \int \sin^{n-2} x \, dx$$

and we have connected the integral of $\sin^n x$ with that of $\sin^{n-2} x$. This integral is particularly simple if the range of integration is $(0, \tfrac{1}{2}\pi)$, for then

$$n \int_0^{\frac{1}{2}\pi} \sin^n x \, dx = (n-1) \int_0^{\frac{1}{2}\pi} \sin^{n-2} x \, dx.$$

By repeated application of this reduction formula we find that $\int_0^{\frac{1}{2}\pi} \sin^n x \, dx$ is

$$\frac{(n-1) \dots 4.2}{n \dots 5.3} \quad \text{or} \quad \frac{(n-1) \dots 3.1}{n \dots 4.2} \frac{\pi}{2}$$

according as *n* is odd or even.

Irrational functions. We take only the simplest cases.

Example.

$$\int \frac{dx}{(x+a)\sqrt{(x+b)}}.$$

The function under the $\sqrt{}$ sign is linear, and the substitution $x+b = u^2$ gives the integral of a rational function of *u*.

Next consider the square root of a quadratic $px^2 + 2qx + r$. The change of variable $u = px + q$ reduces the irrationality to one of the forms

$$\sqrt{(a^2 - x^2)}, \quad \sqrt{(x^2 - a^2)}, \quad \sqrt{(x^2 + a^2)}.$$

The trigonometric or hyperbolic substitution which will get rid of the $\sqrt{}$ is respectively

$$x = a\sin u, \quad x = a\cosh u, \quad x = a\sinh u.$$

Example. $\qquad\qquad \int \sqrt{(8x^2+1)}\, dx.$

The substitution $x = (\sinh u)/2\sqrt{2}$ gives

$$\frac{1}{2\sqrt{2}}\int \cosh^2 u\, du = \frac{1}{4\sqrt{2}}\int (1+\cosh 2u)\, du = \frac{1}{4\sqrt{2}}(u+\sinh u \cosh u),$$

and so, on reverting to x,

$$\frac{1}{4\sqrt{2}} \operatorname{arg sinh} 2x\sqrt{2} + \tfrac{1}{2}x\sqrt{(8x^2+1)}.$$

Thus an integral like

$$\int \sqrt{(ax^2+2bx+c)}\, dx$$

can be expressed in terms of (inverse) trigonometric or hyperbolic functions. If we replace the quadratic under the $\sqrt{}$ sign by a polynomial of higher degree, then we should have to add to our stock of standard functions if we are to express the integral explicitly. (The functions known as elliptic functions would enable us to deal with a cubic or quartic.)

The fact that only the simplest functions are amenable to explicit integration underlines the importance of §§7.12 and 7.13 on approximate methods.

Exercises 7 (b)

Notes on these exercises are given on p. 181.

1. Integrate

$$\frac{1}{(x-2)^2\,(x^2+1)}, \quad \frac{x^4}{(x^2+a^2)\,(x^2+b^2)}, \quad \frac{1}{x^4+a^2x^2+a^4},$$

$$\frac{1}{x^5+1}, \quad \frac{1}{x^6+1}.$$

2. Integrate

$$\frac{x^5}{(x^2+a^2)^4},$$

(i) by substituting $u = x^2+a^2$, (ii) by substituting $x = a\tan\theta$. Verify that the two results agree.

3. *Wallis's product for π.*

Writing $I_n = \displaystyle\int_0^{\frac{1}{2}\pi} \sin^n x\, dx$, prove that I_{2m}/I_{2m+1} lies between 1 and $1+(1/2m)$.

Deduce that

$$\frac{\pi}{2} = \lim_{m\to\infty} \frac{2\,2\,4\,4\,6\,6}{1\,3\,3\,5\,5\,7} \cdots \frac{2m}{2m-1}\frac{2m}{2m+1}.$$

Establish the alternative formula

$$\sqrt{\pi} = \lim_{m\to\infty} \frac{(m!)^2\,2^{2m}}{(2m)!\sqrt{m}}.$$

4. Integrate $\cos^2 3x \sin^3 2x$, $\tan x \sec x$, $\operatorname{cosec}^3 x$.

5. Evaluate $\displaystyle\int_0^\pi \sin 7x \sin nx\, dx$, $\displaystyle\int_0^\pi \sin 7x \sin^3 nx\, dx$.

6. Show how to integrate

$$\frac{a+b\cos x+c\sin x}{p+q\cos x+r\sin x}, \qquad \frac{1}{a\cos^2 x+2b\cos x\sin x+c\sin^2 x}.$$

7. Obtain reduction formulae for the integrals of

$$1/(x^2+1)^n, \quad x^m(\log x)^n, \quad x^n\sqrt{(a^2-x^2)},$$
$$\tan^n x, \quad \sec^n x, \quad 1/(a+b\cos x)^n.$$

8. Find a reduction formula for the integral

$$\int \frac{x^n\,dx}{\sqrt{(ax^2+2bx+c)}}$$

and use it to evaluate

$$\int \frac{x^3\,dx}{\sqrt{(x^2+2x+2)}}.$$

9. Prove that, according as n is an even or odd positive integer,

$$\int_0^\pi \frac{\sin n\theta}{\sin\theta}\,d\theta = 0 \quad\text{or}\quad \pi.$$

If n is a positive integer, evaluate

$$\int_0^\pi \frac{\sin^2 n\theta}{\sin^2\theta}\,d\theta.$$

10. If the polynomial $P_n(x)$ is defined by

$$P_n(x) = \frac{1}{2^n n!}\left(\frac{d}{dx}\right)^n (x^2-1)^n,$$

prove that
  (i) if $Q(x)$ is a polynomial of degree less than n,

$$\int_{-1}^1 P_n(x)\,Q(x)\,dx = 0,$$

  (ii) $\displaystyle\int_{-1}^1 P_m(x)\,P_n(x)\,dx$ is 0 if $m \neq n$ and is $2/(2n+1)$ if $m = n$.

11. Prove that

$$\int \frac{ax^2+2bx+c}{(Ax^2+2Bx+C)^2}\,dx$$

is a rational function of x if and only if $AC-B^2$ or $aC+cA-2bB$ is zero.

7.9. The constant π

We found in §6.8 that $\cos x$ and $\sin x$, defined by their series, are periodic, having a period which we denoted by 2ϖ. We still have to show that ϖ is the same number as the π which presents itself in the geometry of the circle.

Take a circle with centre O and radius a. We recall the argument proving its area to be πa^2: inscribe a regular polygon in the circle; its area is $\frac{1}{2}lp$, where l is the perimeter and p is the perpendicular from O to a side. Similarly a circumscribed regular polygon has area $\frac{1}{2}l_1 a$, where l_1 is its perimeter.

So, if A is the area of the circle,

$$\tfrac{1}{2}lp < A < \tfrac{1}{2}l_1 a.$$

As the number of sides of the polygon tends to infinity, p tends to a, while l_1 and l both tend to the length of the circumference, namely $2\pi a$. So
$$A = \pi a^2.$$

We now collate this with the area found by integration, using the trigonometric functions (defined by series) to evaluate the integral

$$\tfrac{1}{2}A = \text{area of the semi-circle for which } y > 0$$

$$= \int_{-1}^{1} \sqrt{(a^2 - x^2)}\, dx = -\int_{1}^{-1} \sqrt{(a^2 - x^2)}\, dx.$$

Put $x = a\cos\theta$. This gives

$$\tfrac{1}{2}A = \int_{0}^{\varpi} a^2 \sin^2\theta\, d\theta = \tfrac{1}{2}a^2 \int_{0}^{\varpi} (1 - \cos 2\theta)\, d\theta = \tfrac{1}{2}\varpi a^2.$$

We have thus shown that ϖ and π are the same number.

We may conveniently insert here a note on the numerical calculation of π. The easiest way is to use the power series for the inverse tangent.

Theorem 7.9. *If* $-1 \leqslant x \leqslant 1$,

$$\operatorname{arc\,tan} x = x - \tfrac{1}{3}x^3 + \tfrac{1}{5}x^5 - \dots.$$

Proof. By theorems 6.92 and 7.63,

$$\operatorname{arc\,tan} x = \int_{0}^{x} \frac{dt}{1 + t^2}.$$

Now $\dfrac{1}{1+t^2} = 1 - t^2 + t^4 - \ldots + (-1)^{m-1} t^{2m-2} + \dfrac{(-1)^m t^{2m}}{1+t^2}.$

Hence

$$\arctan x = x - \frac{x^3}{3} + \ldots + (-1)^{m-1} \frac{x^{2m-1}}{2m-1} + (-1)^m R_m,$$

where $$R_m = \int_0^x \frac{t^{2m}}{1+t^2}\, dt.$$

If $0 \leqslant x \leqslant 1$,

$$0 \leqslant R_m \leqslant \int_0^x t^{2m}\, dt = \frac{x^{2m+1}}{2m+1} \leqslant \frac{1}{2m+1}$$

and so $R_m \to 0$ as $m \to \infty$. Similarly, if $-1 \leqslant x \leqslant 0$, again $R_m \to 0$. ∎

Putting $x = 1$ in the result of theorem 7.9, we have

$$\tfrac{1}{4}\pi = 1 - \tfrac{1}{3} + \tfrac{1}{5} - \ldots.$$

This gives a means of calculating π, but the series converges too slowly to be useful for numerical work. The following simple relations, which the reader can verify from the addition formulae for the tangent, lead to series which converge more rapidly

$$\tfrac{1}{4}\pi = \arctan\tfrac{1}{2} + \arctan\tfrac{1}{3},$$

$$\tfrac{1}{4}\pi = 4\arctan\tfrac{1}{5} - \arctan\tfrac{1}{239}.$$

7.10. Infinite integrals

Integrals over an infinite interval.

We have, taking a simple example,

$$\int_1^X \frac{dx}{x^2} = 1 - \frac{1}{X}.$$

As $X \to \infty$, the right-hand side tends to the limit 1. A suitable notation to express this fact is

$$\int_1^\infty \frac{dx}{x^2} = 1.$$

Geometrically, the area between the curve $y = 1/x^2$, its asymptote the x-axis and the line $x = 1$ is finite.

Definition. If, as $X \to \infty$,

$$\int_a^X f(x)\, dx \to l,$$

we say that $\int_a^\infty f(x)\, dx$ *exists*, or *converges*, and that its value is l.

If $\int_a^X f(x)\, dx$ exists for all values of X greater than a, but does not tend to a finite limit as $X \to \infty$, we say that $\int_a^\infty f(x)\, dx$ diverges. (It is possible to be more precise—as indicated in §2.9—and separate out divergence to $+\infty$ or to $-\infty$ and finite or infinite oscillation.)

A similar definition applies to

$$\int_{-\infty}^a f(x)\, dx.$$

If $\qquad \int_a^\infty f(x)\, dx = l_1 \quad$ and $\quad \int_{-\infty}^a f(x)\, dx = l_2,$

we write $\qquad \int_{-\infty}^\infty f(x)\, dx = l_1 + l_2.$

It is easy to see that the value of the last integral is independent of the particular value of a.

Theorem 7.10. $\int_1^\infty \dfrac{dx}{x^k}$ *converges if and only if* $k > 1$.

Proof. If $k \neq 1$,

$$\int_1^X \frac{dx}{x^k} = \frac{X^{1-k} - 1}{1 - k}.$$

The limit of the right-hand side is finite if and only if $k > 1$.

If $k = 1$,

$$\int_1^X \frac{dx}{x} = \log X,$$

which tends to infinity as $X \to \infty$. ∎

Note. Any number greater than 0 would serve instead of 1 as the lower limit of integration.

Integrals of unbounded functions. If $\delta > 0$, the function $1/\sqrt{x}$ is continuous in $(\delta, 1)$ and

$$\int_\delta^1 \frac{dx}{\sqrt{x}} = 2 - 2\sqrt{\delta}.$$

Since $1/\sqrt{x}$ is unbounded in the interval $(0, 1)$, the construction of approximative sums cannot be applied directly to define its integral over $(0, 1)$. We use instead the result of making $\delta \to 0$ in the above equation and define

$$\int_0^1 \frac{dx}{\sqrt{x}}$$

to be

$$\lim_{\delta \to 0} \int_\delta^1 \frac{dx}{\sqrt{x}},$$

that is to say, 2.

Such an integral is called an infinite integral of the second kind.

The reader will be able to frame a definition for a general function, following the discussion of $1/\sqrt{x}$.

Exercises 7 (c)

Notes on these exercises are given on p. 181.

1. Evaluate the integrals

$$\int_0^\infty \frac{x\,dx}{x^3+x^2+x+1}, \qquad \int_{\sqrt 2}^\infty \frac{dx}{(x^2-1)\sqrt{(x^2+1)}}.$$

2. Prove that, if $0 < \alpha < \pi$,

$$\int_0^\infty \frac{dx}{x^2+2x\cos\alpha+1} = \frac{\alpha}{\sin\alpha}.$$

3. Prove that

$$\int_0^\infty x^n\,e^{-x}\,dx = n!$$

4. Evaluate

$$\int_0^\infty e^{-ax}\cos bx\,dx \quad (a > 0),$$

$$\int_{-\infty}^\infty \operatorname{sech} ax\,dx.$$

5. Prove that

$$\int_a^b \frac{dx}{\sqrt{\{(x-a)(b-x)\}}} \quad (b > a)$$

exists. Calculate its value by two different substitutions,

(i) $x = a\cos^2\theta + b\sin^2\theta$, \qquad (ii) $(b-x)/(x-a) = u^2$.

6. Evaluate

$$I = \int_{-\delta}^\delta \frac{1-r\cos\theta}{1-2r\cos\theta+r^2}\,d\theta \quad (0 < \delta < \pi)$$

when (i) $0 < r < 1$, (ii) $r > 1$.

Prove that, δ being fixed, I tends to one limit as $r \to 1$ through values less than 1, and to a different limit as $r \to 1$ through values greater than 1. Show, also, that neither limit is equal to the value of I when $r = 1$.

7.11. Series and integrals

There are close analogues between the convergence properties of infinite series and those of infinite integrals. In §2.12 some elementary theorems about convergence of series were proved. The reader should decide what are the corresponding statements about integrals. As an illustration we state the analogue of (6), leaving the proof to the reader.

If, for every $x \geqslant a$,

(1) $f(x) \geqslant 0, g(x) \geqslant 0$;

(2) $f(x) \leqslant Kg(x)$, *where K is a constant;*

(3) $\displaystyle\int_a^\infty g(x)\,dx$ *converges;*

then $\displaystyle\int_a^\infty f(x)\,dx$ *converges. Also* $\displaystyle\int_a^\infty f(x)\,dx \leqslant K\int_a^\infty g(x)\,dx$.

Some care is necessary in framing the analogues. We know from (3) of §2.12 that, if Σu_n converges, then $u_n \to 0$. From this we might expect that, if $\displaystyle\int_a^\infty f(x)\,dx$ converges, then

$$f(x) \to 0 \quad \text{as} \quad x \to \infty.$$

But the following illustration shows that this is *not* the correct conclusion.

Define a function f whose graph consists of the segments of straight lines shown in figure 4.

The height of the peak at each value $x = n$ is 1. The breadth of the triangular base with centre n is $2/(n+1)^2$. f is zero at points not on the sides of one of the triangles. The area of the triangle above $x = n$ is $1/(n+1)^2$, and so

$$\int_0^X f(x)\,dx < \sum_1^\infty \frac{1}{(n+1)^2} \quad \text{(all } X\text{)},$$

showing that $\displaystyle\int_0^\infty f(x)\,dx$ converges. But $f(x)$ does not tend to 0 as $x \to \infty$. The conclusion that we *can* draw is that $\displaystyle\int_n^{n+1} f(x)\,dx$ tends to 0 as $n \to \infty$.

An infinite integral, though it is analogous to an infinite series, is nevertheless a less simple concept. The sum of an infinite series is the result of a single limiting operation ($\lim s_n$ as $n \to \infty$). An integral over a finite range, $\int_0^X f(x)\,dx$ is already a limit (the limit of sums $\Sigma f(\xi_r)\delta_r$). An integral over an infinite range $\int_0^\infty f(x)\,dx$ is thus a limit of a limit, that is to say, a repeated limit.

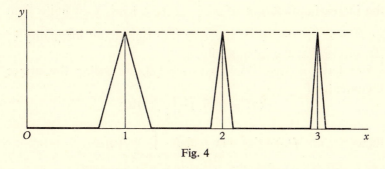

Fig. 4

The following simple and important theorem gives a close connection between the sum of a series of positive decreasing terms and an associated integral.

Theorem 7.11. (*The Maclaurin–Cauchy integral theorem.*) *Let $f(x)$ be, for $x \geqslant 1$, a positive decreasing function of x. Then*

(1) *the integral $\int_1^\infty f(x)\,dx$ and the series $\sum_1^\infty f(n)$ both converge or both diverge;*

(2) *as $n \to \infty$,*

$$\sum_{r=1}^n f(r) - \int_1^n f(x)\,dx$$

tends to a limit l such that $0 \leqslant l \leqslant f(1)$.

Proof. Since $f(x)$ is decreasing, its integrability in any finite interval $(1, X)$ follows from theorem 7.42.

If $n-1 \leqslant x \leqslant n$, we have

$$f(n-1) \geqslant f(x) \geqslant f(n).$$

Integration from $n-1$ to n then gives

$$f(n-1) \geqslant \int_{n-1}^{n} f(x)\,dx \geqslant f(n). \tag{A}$$

Add these inequalities for the intervals $(1, 2)$, $(2, 3)$, ... $(n-1, n)$ and we have

$$\sum_{1}^{n-1} f(r) \geqslant \int_{1}^{n} f(x)\,dx \geqslant \sum_{2}^{n} f(r). \tag{B}$$

If now the series converges, the left-hand inequality shows that the increasing function of X, $\int_{1}^{X} f(x)\,dx$, tends to a finite limit as $X \to \infty$. If the series diverges, the right-hand inequality of (B) shows that the integral diverges.

We have proved (1). To prove (2), we refine the above argument. If

$$\phi(n) = \sum_{1}^{n} f(r) - \int_{1}^{n} f(x)\,dx,$$

then

$$\phi(n) - \phi(n-1) = f(n) - \int_{n-1}^{n} f(x)\,dx$$

$$\leqslant 0 \quad \text{(from A)}.$$

Also, from (B), $\qquad 0 \leqslant \phi(n) \leqslant f(1).$

Therefore the decreasing function $\phi(n)$ tends to a limit l which satisfies

$$0 \leqslant l \leqslant f(1). \ \blacksquare$$

For many functions f it is possible to calculate the integral $\int f(x)\,dx$ but impossible to obtain an explicit sum for the series $\Sigma f(n)$. The Maclaurin–Cauchy theorem is useful for such series. Putting $f(x) = 1/x$ in theorem 7.11 (2), we have the important corollary.

Corollary (*Euler's constant*). *As* $n \to \infty$,

$$1 + \frac{1}{2} + \frac{1}{3} + \dots + \frac{1}{n} - \log n$$

tends to a finite limit γ, *where* $0 < \gamma < 1$.

Euler's constant, γ, is of frequent occurrence in analysis. Its value is $0 \cdot 577 \dots$.

Exercises 7 (d)

Notes on these exercises are given on p. 181.

1. Prove that

$$\log \frac{n}{n-1} - \frac{1}{n} = \int_0^1 \frac{t}{(n-t)n}\, dt \quad (n = 2, 3, \ldots).$$

Denoting either of these expressions by u_n, prove that

$$0 < u_n < \frac{1}{2(n-1)n},$$

and that the series $\sum\limits_{n=2}^{\infty} u_n$ converges to a sum U satisfying $0 < U < \frac{1}{2}$.

Deduce that $\frac{1}{2} < \gamma < 1$.

(The corollary showed only that $0 < \gamma < 1$.)

2. Obtain the limits as $n \to \infty$ of

$$\frac{1}{n+1} + \frac{1}{n+2} + \ldots + \frac{1}{2n},$$

$$\frac{1}{n+1} - \frac{1}{n+2} + \ldots + \frac{(-1)^{n-1}}{2n}.$$

3. Prove that

$$\sum_1^{\infty} \frac{1}{n^2+1} < \frac{1}{2} + \frac{1}{4}\pi.$$

4. Prove that, if $0 \geqslant k > -1$,

$$1^k + 2^k + \ldots + n^k - \frac{n^{k+1}}{k+1}$$

tends to a finite limit as $n \to \infty$.

Deduce that

$$\frac{1^k + 2^k + \ldots + n^k}{n^{k+1}} \to \frac{1}{k+1}.$$

5. Prove that, as $k \to 0+$,

$$k \sum_1^{\infty} \frac{1}{n^{1+k}} \to 1.$$

6. Prove that

$$\sum_2^{\infty} \frac{1}{n(\log n)^k}$$

converges if $k > 1$ and diverges if $k \leqslant 1$.

7. Prove that

$$\int_1^n \log x\, dx < \sum_{r=2}^{n} \log r < \int_1^n \log x\, dx + \log n.$$

Investigate the limit, as $n \to \infty$, of $(n!)^{1/n}/n$.

8. Verify the statement in the numerical illustration at the end of theorem 2.11 that the sum of 10^8 terms of $\Sigma(1/n)$ is less than 20.

9. Discuss the convergence of

$$\text{(i)} \ \Sigma \frac{1}{n^p (\log n)^q}, \quad \text{(ii)} \ \Sigma \frac{1}{n(\log \log n)^q}.$$

7.12. Approximations to definite integrals

At the end of §7.8 we remarked that we are not able to evaluate exactly an integral such as

$$\int_0^1 \frac{dx}{\sqrt{(x^3+1)}},$$

because the integrand containing the square root of a cubic polynomial is not the derivative of any finite combination of our standard functions. You will have to take on trust the fact that we cannot find an indefinite integral explicitly; proofs of impossibility are difficult and far outside the scope of this book. The failure of the usual devices like substitution or integration by parts will incline you to believe that there is no indefinite integral that a search will bring to light. The problem therefore presents itself of obtaining an approximate numerical value of the definite integral.

Another class of integrand for which approximation is forced on us is one which is not specified by an analytical formula at all but, say, by a recording pen attached to an instrument measuring some physical quantity.

Suppose then that we seek an approximation to an integral $\int_a^b f(x)\,dx$, which we cannot evaluate exactly. If we take functions g, h, such that

$$g(x) \geqslant f(x) \geqslant h(x) \quad (a \leqslant x \leqslant b)$$

then the integrals over (a, b) of g and h give approximations from above and from below to the integral of f. Applications of this method are given in the following examples.

Example 1. Consider the integral mentioned at the beginning of this section

$$I = \int_0^1 \frac{dx}{\sqrt{(x^3+1)}}.$$

The $\sqrt{}$ shows that Schwarz's inequality of §7.5 (8) will provide an approximation from above

$$I^2 < \int_0^1 \frac{dx}{x^3+1}$$

$$= \left[\tfrac{1}{3} \log (x+1) - \tfrac{1}{6} \log (x^2-x+1) + \frac{1}{\sqrt{3}} \arctan \frac{2x-1}{\sqrt{3}} \right]_0^1$$

$$= \tfrac{1}{3} \log 2 + \frac{\pi}{3\sqrt{3}} < 0.836,$$

the indefinite integral being obtained by the method of §7.8.

This gives $I < 0.915$.

To find an approximation from below, observe that $x^2 > x^3$ for $0 < x < 1$, and so

$$I > \int_0^1 \frac{dx}{\sqrt{(x^2+1)}} = \Big[\log\{x+\sqrt{(x^2+1)}\}\Big]_0^1 = \log(1+\sqrt{2})$$
$$> 0.896.$$

Methods of obtaining closer approximations will be given in §7.13.

Example 2. Approximate to

$$I = \int_0^{\frac{1}{2}\pi} \sqrt{(\sin x)}\, dx.$$

(The indefinite integral is not expressible by the special functions included in chapter 6.)

(i) $\sqrt{(\sin x)} > \sin x$ gives $I > 1$.

(ii) We have $2x/\pi < \sin x < x \ (0 < x < \frac{1}{2}\pi)$. Integrating the square roots, we find

$$1.047 < I < 1.31.$$

(iii) Schwarz's inequality (as in example 1) yields

$$I < \sqrt{(\tfrac{1}{2}\pi)} < 1.254.$$

7.13. Approximations by subdivision. Simpson's rule

From the definition of the integral, it is natural to carry out approximations by subdividing the range of integration and, if possible, keeping control over the errors that may be incurred in the separate parts. We give some simple and useful methods.

The trapezium method. A first approximation to

$$I = \int_c^d f(x)\, dx$$

is

$$T = \tfrac{1}{2}(d-c)\{f(c)+f(d)\}.$$

If we make the assumption that f has a bounded second derivative, we can obtain an upper bound for the error in this estimate, as follows.

Theorem 7.131. *If* $|f''(x)| \leqslant M$, *then*

$$|I-T| \leqslant \tfrac{1}{12}M(d-c)^3.$$

Proof. For convenience, we can take $c = 0$, $d = h$. The argument is like that of theorem 7.7. Write

$$\phi(h) = \int_0^h t(h-t)f''(t)\, dt.$$

Integrate by parts twice and we have successively

$$\phi(h) = \int_0^h (2t - h) f'(t) \, dt$$

$$= h\{f(h) + f(0)\} - 2\int_0^h f(t) \, dt$$

and so

$$\int_0^h f(x) \, dx = \tfrac{1}{2}h\{f(0) + f(h)\} + R,$$

where

$$|R| \leqslant \tfrac{1}{2}M \int_0^h t(h - t) \, dt = \tfrac{1}{12}Mh^3. \quad \blacksquare$$

To approximate to the integral of $f(x)$ over (a, b), we may divide the interval into n equal parts, where $b - a = nh$, and apply the trapezium rule to each part. The approximate value is then

$$h\{\tfrac{1}{2}f(a) + f(a + h) + f(a + 2h) + \dots + \tfrac{1}{2}f(a + nh)\},$$

the first and last terms having the coefficient $\tfrac{1}{2}$. If, moreover, $|f''(x)| \leqslant M$, then the error is at most

$$\tfrac{1}{12}Mnh^3 \quad \text{or} \quad \tfrac{1}{12}M \frac{(b - a)^3}{n^2}.$$

Simpson's rule. This method, based on the idea of approximating to the curve $y = f(x)$ by a parabola drawn through three of its points (instead of by a straight line through two points) is likely to give a much closer estimate of the integral.

The Simpson approximation to

$$I = \int_c^d f(x) \, dx$$

is

$$S = \tfrac{1}{6}(d - c)\{f(c) + 4f[\tfrac{1}{2}(c + d)] + f(d)\}.$$

To obtain an upper bound for the error we shall assume that $f(x)$ has a bounded fourth derivative.

Lemma. If $y = p(x) = lx^2 + mx + n$ is the parabola (with axis parallel to Oy) drawn through the points of $y = f(x)$ for which $x = -h, 0, h$, then

$$\int_{-h}^h p(x) \, dx = \tfrac{1}{3}h\{f(-h) + 4f(0) + f(h)\}.$$

The proof of the lemma is left to the reader.

Define now, as a measure of the 'error',

$$E(h) = \int_{-h}^{h} f(x)\,dx - \tfrac{1}{3}h\{f(-h) + 4f(0) + f(h)\}.$$

Theorem 7.132. *If* $|f^{(4)}(x)| \leqslant M$, *then* $|E(h)| \leqslant \tfrac{1}{90}Mh^5$.
Proof. Let $0 \leqslant x \leqslant h$. Then

$$
\begin{aligned}
E'(x) &= f(x) + f(-x) - \tfrac{1}{3}\{f(-x) + 4f(0) + f(x)\} \\
&\qquad\qquad - \tfrac{1}{3}x\{f'(x) - f'(-x)\} \\
&= \tfrac{2}{3}f(x) - \tfrac{4}{3}f(0) + \tfrac{2}{3}f(-x) - \tfrac{1}{3}x\{f'(x) - f'(-x)\}.
\end{aligned}
$$

$$E''(x) = \tfrac{1}{3}f'(x) - \tfrac{1}{3}f'(-x) - \tfrac{1}{3}x\{f''(x) + f''(-x)\},$$

$$
\begin{aligned}
E'''(x) &= -\tfrac{1}{3}x\{f'''(x) - f'''(-x)\} \\
&= -\tfrac{2}{3}x^2 f^{(4)}(\xi), \quad \text{where} \quad -x < \xi < x,
\end{aligned}
$$

by the mean value theorem. So

$$-\tfrac{2}{3}Mx^2 \leqslant E'''(x) \leqslant \tfrac{2}{3}Mx^2.$$

Integrating from 0 to x, and noting that $E''(0) = 0$, we have

$$-\tfrac{2}{9}Mx^3 \leqslant E''(x) \leqslant \tfrac{2}{9}Mx^3.$$

Integrating twice more, and using $E'(0) = E(0) = 0$, we have

$$-\tfrac{1}{18}Mx^4 \leqslant E'(x) \leqslant \tfrac{1}{18}Mx^4 \quad (0 \leqslant x \leqslant h)$$

and finally $\quad -\tfrac{1}{90}Mh^5 \leqslant E(h) \leqslant \tfrac{1}{90}Mh^5.$ ∎

In practice, to approximate to $\int_a^b f(x)\,dx$, divide the interval (a, b) into $2n$ equal parts, where $b - a = 2nh$. Let the values of $f(x)$ at the $2n + 1$ end-points of the subintervals be

$$y_0, y_1, \ldots, y_{2n}.$$

Applying the method of the lemma to the n sets of two adjacent subintervals, we have the approximate value

$$\tfrac{1}{3}h\{(y_0 + y_{2n}) + 4(y_1 + y_3 + \ldots + y_{2n-1}) + 2(y_2 + y_4 + \ldots + y_{2n-2})\}.$$

If, further, $|f^{(4)}(x)| \leqslant M$, theorem 7.132 shows that the error is at most

$$\tfrac{1}{90}Mn\left(\frac{b-a}{2n}\right)^5 = \tfrac{1}{2880}M\frac{(b-a)^5}{n^4}.$$

Exercises 7 (e) (*Approximations*)

Notes on these exercises are given on p. 181.

1. Prove that, if $f(x) = (1 + \frac{1}{2}x - \frac{1}{10}x^2)/\sqrt{(1+x)}$, then $1 \leqslant f(x) < 1 \cdot 0008$ in the interval $0 \leqslant x \leqslant \frac{1}{2}$. Hence evaluate the integral

$$\int_0^{\frac{1}{2}} \frac{1 + \frac{1}{2}x - \frac{1}{10}x^2}{\sqrt{(1-x^2)}}\, dx$$

correct to three places of decimals ($\sqrt 2 = 1 \cdot 4142...$).

2. Prove that

$$\int_0^1 \frac{u^4(1-u)^4}{1+u^2}\, du = \tfrac{22}{7} - \pi.$$

Evaluate $\int_0^1 u^4(1-u)^4\, du$ and deduce that

$$\tfrac{22}{7} - \tfrac{1}{1260} > \pi > \tfrac{22}{7} - \tfrac{1}{630}.$$

3. If $\phi(x)$ is polynomial of the fifth degree, prove that

$$\int_0^1 \phi(x)\, dx = \tfrac{1}{18}\{5\phi(\alpha) + 8\phi(\tfrac{1}{2}) + 5\phi(\beta)\},$$

where α and β are the roots of the equation $x^2 - x + \frac{1}{10} = 0$.

4. The function $f(x)$ has a continuous nth derivative for $x \geqslant 0$, and $f(x)$ and its first $n-1$ derivatives vanish for $x = 0$. Show that

$$\int_0^a \frac{(a-x)^n}{n!} f^{(n)}(x)\, dx = \int_0^a f(x)\, dx.$$

Deduce that, if $|f^{(n)}(x)| \leqslant M$ in $0 \leqslant x \leqslant a$, then

$$\left| \int_0^a f(x)\, dx \right| \leqslant \frac{a^{n+1}}{n!} \frac{M}{\sqrt{(2n+1)}}.$$

5. Discuss the common assumption that the Simpson approximation to an integral, obtained by dividing the range (a, b) into $2n$ equal parts, is liable to an error varying as n^{-4}.

If I_1 and I_2 are the approximations found by Simpson's rule when the range is divided into $2n$ and $4n$ parts, respectively, show that, on the above assumption, $(16I_2 - I_1)/15$ is a better approximation.

Apply this to $\int_4^8 \frac{dx}{x}$ with $n = 1$, and compare your result with the true value $0 \cdot 693147...$.

6. (*Stirling's formula for $n!$.*) This formula gives a good approximation to $n!$ when n is large (J. Stirling, 1692–1770).
Prove that, as $n \to \infty$,

$$\phi(n) = \frac{n!}{n^{n+\frac{1}{2}} e^{-n}} \to \sqrt{(2\pi)}.$$

The proof is in two parts: (*a*) prove that $\phi(n) \to$ some constant A, (*b*) prove that $A = \sqrt{(2\pi)}$.

(a) If r is an integer $\geqslant 2$,

(i) $\displaystyle\int_{r-\frac{1}{2}}^{r+\frac{1}{2}} \log x\,dx < \log r, \quad \frac{1}{2}\{\log(r-1)+\log r\} < \int_{r-1}^{r} \log x\,dx.$

(ii) $\displaystyle\int_{\frac{3}{2}}^{n} \log x\,dx < \log(n!) - \frac{1}{2}\log n < \int_{1}^{n} \log x\,dx.$

(iii) If $u_n = \log(n!) - (n+\frac{1}{2})\log n + n$, then

$$u_n > u_{n-1} \quad\text{and}\quad \tfrac{3}{2}(1-\log\tfrac{3}{2}) < u_n < 1.$$

(iv) $\phi(n) \to A$, where $2{\cdot}45 < A < e$.

(b) Apply Wallis's formula (exercise 7 (b), 3) to $\phi(2n)/\{\phi(n)\}^2$.

Exercises 7 (f) (*Miscellaneous*)

Notes on these exercises are given on pp. 181–2.

1. Find the limits, as x tends to 0 through positive values, of

$$\text{(i)}\ \frac{1}{x}\int_{0}^{x} \sqrt{(3t^2+2)}\,dt, \quad \text{(ii)}\ \frac{1}{x^2}\int_{-x}^{x} |t|\,dt.$$

2. If

$$G(x,\xi) = \begin{cases} x(\xi-1) & \text{when}\ \ x \leqslant \xi, \\ \xi(x-1) & \text{when}\ \ \xi \leqslant x, \end{cases}$$

and if $f(x)$ is a continuous function of x in $0 \leqslant x \leqslant 1$ and if

$$g(x) = \int_{0}^{1} f(\xi)\,G(x,\xi)\,d\xi,$$

show that $\qquad\qquad g''(x) = f(x)$

and find $g(0)$ and $g(1)$.

3. If

$$f(p,q) = \int_{0}^{1} x^{p-1}(1-x)^{q-1}dx,$$

where $p \geqslant 1, q \geqslant 1$, show that

$$f(p+1,q) + f(p,q+1) = f(p,q)$$

and $\qquad\qquad qf(p+1,q) = pf(p,q+1).$

Evaluate $f(p,n)$, where n is a positive integer.

4. Let

$$S_r = \int^{\pi} \sin^r\theta\,d\theta \quad (r \geqslant 0), \quad P_r = rS_rS_{r-1} \quad (r \geqslant 1),$$

where r is not necessarily an integer. Prove that

$$\text{(i)}\ P_r = P_{r+1} \quad (r \geqslant 1), \qquad \text{(ii)}\ P_1 = \tfrac{1}{2}\pi,$$

$$\text{(iii)}\ P_r/r \text{ decreases as } r \text{ increases } (r \geqslant 1).$$

Deduce from (i), (ii), (iii) that

$$\text{(iv)}\ \frac{r}{k+1}\frac{\pi}{2} < P_r < \frac{r}{k}\frac{\pi}{2} \quad (1 \leqslant k \leqslant r < k+1;\ k \text{ an integer});$$

and from (i), (iv) that

$$\text{(v)}\ P_r = \tfrac{1}{2}\pi \text{ for all } r \geqslant 1.$$

5. From approximative sums to appropriate integrals, find the limits as $n \to \infty$ of

$$\text{(i)} \quad n \sum_{m=0}^{n-1} \frac{1}{n^2 + m^2},$$

$$\text{(ii)} \quad \frac{\sqrt[n]{\{(n+1)(n+2)\dots(2n)\}}}{n}.$$

6. From the identity

$$1 - a^{2n} = (1 - a^2) \prod_{r=1}^{n-1} \left(1 - 2a \cos \frac{r\pi}{n} + a^2\right),$$

(where Π denotes the products of the factors given by $r = 1, 2, \dots, n-1$), prove that (a being real)

$$\int_0^\pi \log (1 - 2a \cos x + a^2) \, dx$$

is 0 if $|a| < 1$ and is $2\pi \log |a|$ if $|a| > 1$.

7. (*Proof that π is irrational.*) If

$$I_n(\alpha) = \int_{-1}^1 (1 - x^2)^n \cos \alpha x \, dx,$$

prove that, if $n \geqslant 2$,

$$\alpha^2 I_n = 2n(2n-1) I_{n-1} - 4n(n-1) I_{n-2}.$$

Deduce that, for all positive integral values of n,

$$\alpha^{2n+1} I_n(\alpha) = n! \, (P \sin \alpha + Q \cos \alpha),$$

where P and Q are polynomials of degree less than $2n+1$ in α with integral coefficients.

Prove that, if $\frac{1}{2}\pi$ were equal to b/a, where b and a are integers, then

$$b^{2n+1} I_n(\tfrac{1}{2}\pi)/n!$$

would be an integer. By considering large values of n, prove that π is irrational.

8

FUNCTIONS OF SEVERAL VARIABLES

8.1. Functions of x and y

We have applied limiting processes to a function $f(x)$ of a single variable x. This analysis will now be extended to functions which depend on more than one independent real variable. Geometrical language will help towards clearness and brevity. Referred to a pair of rectangular axes, two real numbers correspond to a point in a plane. We now define a function of x and y.

Let E be a set of points P in the (x, y) plane, or, what is equivalent, a set of values of the pair of real numbers (x, y). If rules are given which determine a unique real number z corresponding to each pair (x, y), then z is called a *function* of x and y. The set E is called the *domain* of the function.

We write
$$z = f(x, y)$$
or, commonly
$$z = z(x, y),$$

using, as we can without ambiguity, z to denote both the functional symbol and its numerical value.

Notes. (1) Hitherto the letter z has conveniently denoted a complex number $x + yi$. In this chapter the usage is different and corresponds with that of three-dimensional analytical geometry referred to coordinate axes Ox, Oy, Oz.

(2) The notes (1)–(4) of §3.1 are applicable here with the appropriate modifications. The function f is a transformation or mapping of a two-dimensional domain E of points (x, y) into a linear set of values z under the restriction that with a given pair of values (x, y) is associated one and only one value of z.

Illustrations. $z = (1 - x^2 - y^2)^k$ defines a function of (x, y)

(i) if $k = 2$, for all (x, y),

(ii) if $k = -1$, for all (x, y) with the exception of points on the circle $x^2 + y^2 = 1$,

(iii) if $k = \frac{1}{2}$, for points inside or on the circle $x^2 + y^2 = 1$.

The equation $z = f(x, y)$ represents a surface referred to a set of axes Ox, Oy, Oz. There is no graphical representation of the values of a function $f(x, y)$ so simple as the curve $y = f(x)$ which illustrates a function of one variable. A possible representation on a sheet of paper is obtained by drawing the curves

$$f(x, y) = k$$

for suitable values of k. These are the contour lines of the surface $z = f(x, y)$.

Illustration. The contour-lines of

$$z = x^2 + 4y^2$$

are a set of similar ellipses with centre at the origin.

If there are more than two independent variables we can no longer visualize a functional relation graphically. The analytical methods of this chapter apply to a function of any number of variables; we shall usually suppose for simplicity that the number of variables is two or three.

8.2. Limits and continuity

We first clarify the notion of the limit of a function of more than one variable.

Definition. $f(x, y)$ *tends to the limit* l *as* (x, y) *tends to* (a, b) *if, given* ϵ, *there is* δ *such that*

$$|f(x, y) - l| < \epsilon$$

for all x, y *such that*

$$0 < \sqrt{\{(x-a)^2 + (y-b)^2\}} < \delta.$$

The definition expresses the requirement that $f(x, y)$ can be made as near to l as may be demanded by taking (x, y) to be any point inside a sufficiently small circle with centre (a, b). A square with centre (a, b) and side 2δ, or a region of any other shape surrounding (a, b) would serve just as well as the circle in the definition. We shall call a region such as the circle or the square a *neighbourhood* of (a, b).

Definition. $f(x, y)$ *is continuous at* (a, b) *if* $f(x, y)$ *tends to* $f(a, b)$ *as* (x, y) *tends to* (a, b).

It is natural to go on to define continuity of f throughout a domain E of the (x, y) plane to mean continuity at each point of E. The reader should refer to the definitions of continuity of a function of x in a closed interval and in an open interval stated at the end of §3.4. The following definition is sufficient for our needs in this chapter.

Definition. $f(x, y)$ *is continuous in an open rectangular domain* $x_0 < x < x_1, y_0 < y < y_1$ *if it is continuous at each point of the domain.*

If, however, points of the 'boundary' are included in the set, the formal definition would need more care because in more than one dimension the boundary of a domain is more complicated than the boundary of a linear interval which consists just of its two end-points. The general idea (following that of §3.4) is to exclude from consideration values taken by f at any points not in the set E.

It is to be observed that continuity of $f(x, y)$ as a function of the pair of variables (x, y) asserts more than continuity of $f(x, y)$ as a function of either variable singly (keeping the other fixed).

Illustration. Define

$$f(x, y) = \frac{2xy}{x^2 + y^2} \quad \text{if } x, y \text{ are not both } 0,$$

$$f(0, 0) = 0.$$

Then $f(x, y)$ is 0 at all points of Ox and Oy. So $f(x, 0)$ is a continuous function of x for $x = 0$ and $f(0, y)$ is a continuous function of y for $y = 0$. We shall prove that $f(x, y)$ is not a continuous function of (x, y) at $(0, 0)$. This is easily seen by using polar coordinates $x = r \cos \theta$, $y = r \sin \theta$. Then $f(x, y) = \sin 2\theta$ for all values of r except $r = 0$. So, in any circle however small with centre $(0, 0)$, the function assumes all values between -1 and 1 and does not tend to a limit as (x, y) tends to $(0, 0)$.

8.3. Partial differentiation

We may, keeping y constant, differentiate $f(x, y)$ as a function of x, that is to say, we take the limit

$$\lim_{h \to 0} \frac{f(x+h, y) - f(x, y)}{h}.$$

This is called the partial derivative of $f(x, y)$ with regard to x and can be written

$$\frac{\partial f}{\partial x} \quad \text{or} \quad f_x(x, y) \quad \text{or} \quad f_x.$$

Any of these notations applied to a function of two or more variables indicates that every variable, except the variable of differentiation, is to be kept constant.

The first-order derivatives $\partial f/\partial x$ and $\partial f/\partial y$ are functions of (x, y) and may be differentiated again. We have then

$$\frac{\partial^2 f}{\partial x^2} \quad \text{or} \quad f_{xx} \quad \text{and} \quad \frac{\partial^2 f}{\partial y^2} \quad \text{or} \quad f_{yy}$$

and the mixed second-order partial derivatives

$$\frac{\partial}{\partial y}\left(\frac{\partial f}{\partial x}\right) \quad \text{and} \quad \frac{\partial}{\partial x}\left(\frac{\partial f}{\partial y}\right).$$

In the suffix notation, the former of these is $(f_x)_y$ or, by suppressing the brackets, f_{xy}.

It will be found that in all straightforward examples the order in which partial differentiations are carried out does not matter. We shall usually have, in fact,

$$\frac{\partial}{\partial y}\left(\frac{\partial f}{\partial x}\right) = \frac{\partial}{\partial x}\left(\frac{\partial f}{\partial y}\right)$$

and then we can write each of them as

$$\frac{\partial^2 f}{\partial x\,\partial y} \quad \text{(or } f_{xy}\text{)}.$$

Our first theorem gives conditions for this interchangeability of the order of partial differentiations.

Theorem 8.3. *If f_{xy} and f_{yx} are continuous functions of (x, y) at a point (a, b), they are equal there.*

Proof. Since f_{xy} and f_{yx} are continuous at (a, b), there is a square N, centre (a, b), in which they exist. The functions f_x and f_y from which they are derived must exist in N, as must f. For the rest of the proof, h and k are assumed to be small enough for $(a+h, b+k)$ to be in N.

The proof depends on taking increments in the function f corresponding to increments in the variables x and y in turn. If

$$\phi(x, y) = f(x, y+k) - f(x, y)$$

and

$$D = \phi(x+h, y) - \phi(x, y),$$

then $D = f(x+h, y+k) - f(x+h, y) - f(x, y+k) + f(x, y).$

Now D can be built up by taking increments in x and y in the reverse order, for if

$$\psi(x, y) = f(x+h, y) - f(x, y),$$

then $\qquad D = \psi(x, y+k) - \psi(x, y).$

By the mean value theorem (4.61)

$$\phi(a+h, b) - \phi(a, b) = h\phi_x(a+\theta_1 h, b),$$

where θ_1 and all other θ's in the sequel lie between 0 and 1. By the definition of ϕ,

$$\phi_x(a+\theta_1 h, b) = f_x(a+\theta_1 h, b+k) - f_x(a+\theta_1 h, b).$$

Apply the mean value theorem to the right-hand side and we have, with $(x, y) = (a, b)$ in D,

$$D = hk f_{xy}(a+\theta_1 h, b+\theta_2 k).$$

The alternative form of D as the increment of $\psi(a, y)$ over $(b, b+k)$ gives
$$D = hk f_{yx}(a+\theta_3 h, b+\theta_4 k).$$

So, if $h \neq 0$ and $k \neq 0$,

$$f_{xy}(a+\theta_1 h, b+\theta_2 k) = f_{yx}(a+\theta_3 h, b+\theta_4 k).$$

Now let $(h, k) \to (0, 0)$. The continuity of f_{xy} and f_{yx} at (a, b) gives
$$f_{xy}(a, b) = f_{yx}(a, b). \ |$$

Exercises 8 (a)

Notes on these exercises are given on p. 182.

1. Which of the following functions (with the definition suitably completed for $x = 0$, $y = 0$) are continuous at $(0, 0)$?

(i) $\dfrac{(x+y)^2}{x^2+y^2}$, (ii) $\dfrac{xy^2}{x^2+y^4}$, (iii) $\dfrac{x^3+y^3}{x^2+y^2}$.

2. If $x = r\cos\theta$, $y = r\sin\theta$, give the values of the partial derivatives of r and θ with respect to x and y.

Is it true that
$$\left(\frac{\partial r}{\partial x}\right)\left(\frac{\partial x}{\partial r}\right) = 1?$$

Illustrate by a diagram.

8.4. Differentiability

Let us consider what is the most natural extension of the idea of *differentiability* when the number of independent variables exceeds one. If we have a function f of one variable, differentiability at a point means the existence of a tangent line—a linear approximation to the curve $y = f(x)$. On a surface $z = f(x, y)$ an approximation linear in the variables suggests a tangent plane. This leads us to the following definition.

Definition. The function $f(x, y)$ is differentiable *at (a, b) if, for $(a+h, b+k)$ in a neighbourhood of (a, b),*

$$f(a+h, b+k) - f(a, b) = Ah + Bk + \epsilon(|h| + |k|),$$

where A and B do not depend on h or k, and ϵ tends to 0 as $(h, k) \to (0, 0)$.

This means that the plane

$$z - f(a, b) = A(x - a) + B(y - b)$$

is at a distance from the surface $z = f(x, y)$ which is small compared with the displacement of the point $(a+h, b+k)$ from (a, b) and is the tangent plane to the surface.

The 'error-term' in the definition of differentiability which we have written as

$$\epsilon(|h| + |k|)$$

can be put in several different forms, which are easily seen to be equivalent. We could, for instance, say instead

$$\epsilon \sqrt{(h^2 + k^2)}.$$

Again, we could rewrite the condition as

$$f(a+h, b+k) - f(a, b) = \alpha(h, k)h + \beta(h, k)k,$$

where $\alpha(h, k) \to A$ and $\beta(h, k) \to B$ as $(h, k) \to (0, 0)$.

If (in the original definition) we keep $k = 0$ and let h tend to 0, we have

$$A = f_x(a, b),$$

and similarly

$$B = f_y(a, b).$$

It can be seen that differentiability, in the sense of this definition, asserts more than the existence of partial derivatives with respect to x and y. Geometrically, the existence of the partial derivative $f_x(a, b)$ implies that there is a tangent line to the curve which is the section of the surface $z = f(x, y)$ by the plane $y = b$. The existence of a tangent plane to a surface requires more than the existence of tangent lines to two curves which are sections by perpendicular planes. The next theorem shows that if we assume the continuity, and not merely the existence, of f_x and f_y, then the differentiability of f is a consequence.

Theorem 8.4. *If f_x and f_y are continuous at a point (x, y) then f is differentiable at that point.*

Proof. Let h and k be small enough for $(x+h, y+k)$ to lie within a circular neighbourhood of (x, y) in which f_x and f_y exist. Then we have

$$f(x+h, y+k) - f(x, y)$$
$$= \{f(x+h, y+k) - f(x, y+k)\} + \{f(x, y+k) - f(x, y)\}.$$

In the first bracket, only x is changed; in the second, only y. Apply the mean value theorem to each and we obtain

$$hf_x(x+\theta_1 h, y+k) + kf_y(x, y+\theta_2 k).$$

By continuity of f_x and f_y this is equal to

$$h\{f_x(x, y) + \epsilon_1\} + k\{f_y(x, y) + \epsilon_2\},$$

where the ϵ's tend to 0 as $(h, k) \to (0, 0)$, and the definition of differentiability is satisfied.

Exercises 8 (*b*)

Notes on these exercises are given on p. 182.

1. Investigate whether the following functions are differentiable at $(0, 0)$:

(i) $|x^2 - y^2|$,   (ii) $|xy|^k$.

2. If

$$f(x, y) = \begin{cases} xy/\sqrt{(x^2+y^2)} & \text{when} \quad (x, y) \neq (0, 0), \\ 0 & \text{when} \quad (x, y) = (0, 0), \end{cases}$$

investigate for $(0, 0)$, (i) continuity of f; (ii) existence of f_x, f_y; (iii) differentiability of f.

3. Give an example of $f(x, y)$ which is differentiable at $(0, 0)$ and discontinuous at all other points.

8.5. Composite functions

We now extend to functions of two variables the formula

$$\frac{dy}{dx} = \frac{dy}{du}\frac{du}{dx}$$

proved in §4.2 (5).

Theorem 8.5. *If $x = x(t)$ and $y = y(t)$ are differentiable functions of t for a given t, and $z = z(x, y)$ is a differentiable function of (x, y) for the corresponding (x, y), then*

$$z = z\{x(t), y(t)\}$$

is a differentiable function of t and

$$\frac{dz}{dt} = \frac{\partial z}{\partial x}\frac{dx}{dt} + \frac{\partial z}{\partial y}\frac{dy}{dt}.$$

Proof. Let t be changed to $t + \delta t$; let δx and δy be the corresponding changes in x and y.

Then

$$\delta x = \left(\frac{dx}{dt} + \epsilon_1\right)\delta t,$$

$$\delta y = \left(\frac{dy}{dt} + \epsilon_2\right)\delta t,$$

where ϵ_1 and ϵ_2 tend to 0 as $\delta t \to 0$. Writing

$$\delta z = z(x + \delta x, y + \delta y) - z(x, y),$$

we have, since z is differentiable,

$$\delta z = \frac{\partial z}{\partial x}\delta x + \frac{\partial z}{\partial y}\delta y + \epsilon(|\delta x| + |\delta y|),$$

where $\epsilon \to 0$ as $(\delta x, \delta y) \to (0, 0)$.

If δx and δy are both 0 (which may happen for arbitrarily small δt if $dx/dt = dy/dt = 0$), ϵ fails to be determined by the last equation and we define ϵ to be 0.

Substituting for δx and δy in terms of δt, we obtain

$$\delta z = \left(\frac{\partial z}{\partial x}\frac{dx}{dt} + \frac{\partial z}{\partial y}\frac{dy}{dt}\right)\delta t + \eta\,\delta t,$$

where it is easily seen that $\eta \to 0$ as $\delta t \to 0$. ▮

Corollary 1. If, in the theorem, x and y are functions of more than one variable, say

$$x = x(t, u, v), \quad y = y(t, u, v),$$

which possess partial derivatives, then, for each of the variables, we obtain, by keeping the others constant,

$$\frac{\partial z}{\partial t} = \frac{\partial z}{\partial x}\frac{\partial x}{\partial t} + \frac{\partial z}{\partial y}\frac{\partial y}{\partial t}$$

and similar equations with t replaced in turn by u and v.

This much-used rule for differentiating through the intermediate variables is commonly called the chain rule.

Corollary 2. If, in corollary 1, $x(t, u, v)$ and $y(t, u, v)$ are differentiable in the sense of §8.4, then a straightforward adaptation of the proof of theorem 8.5 shows that

$$z\{x(t, u, v), y(t, u, v)\}$$

is a differentiable function of t, u, v.

8.6. Changes of variable. Homogeneous functions

This section contains applications of theorem 8.5.

Example. Express in polar coordinates

$$\frac{\partial^2 V}{\partial x^2} + \frac{\partial^2 V}{\partial y^2},$$

assuming that these second-order derivatives are continuous.

Solution. Corollary 1 of the last theorem gives

$$\frac{\partial V}{\partial r} = \frac{\partial V}{\partial x}\frac{\partial x}{\partial r} + \frac{\partial V}{\partial y}\frac{\partial y}{\partial r} = \frac{\partial V}{\partial x}\cos\theta + \frac{\partial V}{\partial y}\sin\theta,$$

$$\frac{\partial V}{\partial \theta} = \frac{\partial V}{\partial x}\frac{\partial x}{\partial \theta} + \frac{\partial V}{\partial y}\frac{\partial y}{\partial \theta} = \frac{\partial V}{\partial x}(-r\sin\theta) + \frac{\partial V}{\partial y}r\cos\theta.$$

At this stage we have on the right-hand sides terms containing both (x, y) and (r, θ). To keep the cartesian and polar variables separate, it is advisable to solve for $\partial V/\partial x$ and $\partial V/\partial y$. We find (if $r \neq 0$)

$$\frac{\partial V}{\partial x} = \cos\theta\,\frac{\partial V}{\partial r} - \frac{\sin\theta}{r}\frac{\partial V}{\partial \theta},$$

$$\frac{\partial V}{\partial y} = \sin\theta\,\frac{\partial V}{\partial r} + \frac{\cos\theta}{r}\frac{\partial V}{\partial \theta}.$$

These equations enable us to convert the operators $\partial/\partial x$ and $\partial/\partial y$ acting on a function into operators involving $\partial/\partial r$ and $\partial/\partial\theta$. Thus

$$\frac{\partial^2 V}{\partial x^2} = \left(\cos\theta\,\frac{\partial}{\partial r} - \frac{\sin\theta}{r}\,\frac{\partial}{\partial\theta}\right)\left(\cos\theta\,\frac{\partial V}{\partial r} - \frac{\sin\theta}{r}\,\frac{\partial V}{\partial\theta}\right)$$

$$= \cos\theta\left\{\cos\theta\,\frac{\partial^2 V}{\partial r^2} - \frac{\sin\theta}{r}\,\frac{\partial^2 V}{\partial r\,\partial\theta} + \frac{\sin\theta}{r^2}\,\frac{\partial V}{\partial\theta}\right\}$$

$$- \frac{\sin\theta}{r}\left\{\cos\theta\,\frac{\partial^2 V}{\partial r\,\partial\theta} - \sin\theta\,\frac{\partial V}{\partial r} - \frac{\sin\theta}{r}\,\frac{\partial^2 V}{\partial\theta^2} - \frac{\cos\theta}{r}\,\frac{\partial V}{\partial\theta}\right\}.$$

There is a similar expression for $\partial^2 V/\partial y^2$, which the reader should write down. Adding, he will find that

$$\frac{\partial^2 V}{\partial x^2} + \frac{\partial^2 V}{\partial y^2} = \frac{\partial^2 V}{\partial r^2} + \frac{1}{r}\,\frac{\partial V}{\partial r} + \frac{1}{r^2}\,\frac{\partial^2 V}{\partial\theta^2}. \quad \blacksquare$$

We now turn to homogeneous functions as a further illustration of theorem 8.5. The elementary notion of a homogeneous function is a polynomial in the variables, all the terms being of the same degree. For instance, $2x^3 - x^2 y + 4y^3$ is homogeneous of degree 3. The more general definition which follows allows, say,

$$\frac{(x^3 + 2y^3)^{\frac{1}{2}}}{x + 3y}$$

to be homogeneous of degree $\frac{1}{2}$.

Definition. $f(x, y)$ is *homogeneous of degree h* if, for every positive t and all x, y for which f is defined,

$$f(tx, ty) = t^h f(x, y).$$

Theorem 8.6. *A necessary and sufficient condition that the differentiable function $f(x, y)$ should be homogeneous of degree h is that*

$$xf_x + yf_y = hf$$

for all x, y.

Proof. We prove the necessity, which dates from Euler. Putting $\xi = tx$, $\eta = ty$, we have

$$f(\xi, \eta) = t^h f(x, y).$$

Keeping x, y constant, differentiate with regard to t. Theorem 8.5 gives

$$\frac{\partial f}{\partial \xi}\frac{d\xi}{dt} + \frac{\partial f}{\partial \eta}\frac{d\eta}{dt} = ht^{h-1}f(x, y).$$

Put $h = 1$ and we have the required condition

$$f_x x + f_y y = hf.$$

To prove the sufficiency of this condition, take, as before, $\xi = tx$, $\eta = ty$. Keeping x, y constant, we have

$$\frac{d}{dt}f(\xi, \eta) = x\frac{\partial f}{\partial \xi} + y\frac{\partial f}{\partial \eta} = \frac{1}{t}\left(\xi\frac{\partial f}{\partial \xi} + \eta\frac{\partial f}{\partial \eta}\right),$$

and, by hypothesis, the last expression is $hf(\xi, \eta)/t$. So, if $v = f(\xi, \eta)$,

$$\frac{1}{v}\frac{dv}{dt} = \frac{h}{t}.$$

Therefore $v = At^h$, where A is independent of t. So

$$f(tx, ty) = At^h = t^h f(x, y),$$

by putting $t = 1$. ∎

Exercises 8 (c)

Notes on these exercises are given on p. 182.

1. (*The Jacobian.*) Let u, v be differentiable functions of x, y. The determinant

$$\begin{vmatrix} u_x & u_y \\ v_x & v_y \end{vmatrix}$$

written shortly as

$$\frac{\partial(u, v)}{\partial(x, y)}$$

is called the Jacobian of the transformation from (x, y) to (u, v). Its role corresponds to that of the derivative du/dx for a function $u(x)$ of a single variable. Prove the following properties which illustrate this role.

   (i) If a relation $\phi(u, v) = 0$ holds for all (x, y) in a domain D, then

$$\frac{\partial(u, v)}{\partial(x, y)} = 0$$

in D. (The converse is true but more difficult to prove.)

   (ii) If x, y are differentiable functions of s, t, then

$$\frac{\partial(u, v)}{\partial(s, t)} = \frac{\partial(u, v)}{\partial(x, y)}\frac{\partial(x, y)}{\partial(s, t)}.$$

All this may be extended to n functions of n variables.

2. A function $f(x)$ is defined by the properties $f'(x) = 1/(1+x^2)$ and $f(0) = 0$. Use **1** (i) (the converse) to show that there is a functional equation

$$f(x)+f(y) = f\left(\frac{x+y}{1-xy}\right).$$

3. The area Δ of a triangle is found from measurements of a, B, C. Prove that the error in the calculated value of Δ due to small errors $\delta a, \delta B, \delta C$ is given approximately by

$$\frac{\delta\Delta}{\Delta} = 2\frac{\delta a}{a} + \frac{c}{a\sin B}\delta B + \frac{b}{a\sin C}\delta C.$$

4. The sides a, b, c of a triangle are measured with a possible percentage error ϵ and the area is calculated. Prove that the possible percentage error in the area is approximately 2ϵ or $2\epsilon \cot B \cot C$ according as the triangle is acute-angled or obtuse-angled at A.

5. (*The wave-equation.*) Prove that, by the transformations $u = x-ct$, $v = x+ct$, the partial differential equation

$$\frac{\partial^2 y}{\partial t^2} = c^2\frac{\partial^2 y}{\partial x^2}$$

reduces to

$$\frac{\partial^2 y}{\partial u\,\partial v} = 0,$$

and hence solve the equation.

6. If $f(x, y)$ is transformed into $g(u, v)$ by the transformation

$$x = u^2-v^2, \quad y = 2uv,$$

prove that

$$\frac{\partial^2 f}{\partial x^2} + \frac{\partial^2 f}{\partial y^2} = \frac{1}{4(u^2+v^2)}\left(\frac{\partial^2 g}{\partial u^2} + \frac{\partial^2 g}{\partial v^2}\right).$$

Find the most general function $f(x, y)$ which satisfies

$$\frac{\partial^2 f}{\partial x^2} + \frac{\partial^2 f}{\partial y^2} = 0$$

and is a function of $x + \sqrt{(x^2+y^2)}$ only.

7. Prove that, if r and θ are polar coordinates and $\lambda = \log r$, the equation

$$\frac{\partial^2 V}{\partial x^2} + \frac{\partial^2 V}{\partial y^2} = 0$$

becomes

$$\frac{\partial^2 V}{\partial \lambda^2} + \frac{\partial^2 V}{\partial \theta^2} = 0.$$

8. For all positive values of λ, the function g satisfies the condition that

$$g(\lambda^r x, \lambda^s y) = \lambda^n g(x, y),$$

where r, s, n are positive integers. Prove that

$$rx\frac{\partial g}{\partial x} + sy\frac{\partial g}{\partial y} = ng.$$

9. The function $F(u, v)$ becomes $f(x, y)$ when the substitution

$$u = x^3 - 3xy^2,$$
$$v = 3x^2y - y^3$$

is made. Prove that

$$\frac{1}{9(x^2 + y^2)^2} \left\{ \frac{\partial^2 f}{\partial x^2} + \frac{\partial^2 f}{\partial y^2} \right\} = \frac{\partial^2 F}{\partial u^2} + \frac{\partial^2 F}{\partial v^2}.$$

8.7. Taylor's theorem

We extend to a function of two variables the nth order mean value theorem proved on page 81.

Theorem 8.7. *Suppose that the partial derivatives of order n of* $f(x, y)$ *are continuous in a neighbourhood of* (a, b) *which contains the line joining* (a, b) *to* $(a+h, b+k)$. *Then*

$$f(a+h, b+k) = f(a, b) + \left(h \frac{\partial}{\partial a} + k \frac{\partial}{\partial b} \right) f(a, b) + \dots$$

$$+ \frac{1}{(n-1)!} \left(h \frac{\partial}{\partial a} + k \frac{\partial}{\partial b} \right)^{n-1} f(a, b)$$

$$+ \frac{1}{n!} \left(h \frac{\partial}{\partial a} + k \frac{\partial}{\partial b} \right)^{n} f(a+\theta h, b+\theta k), \quad \text{where } 0 < \theta < 1.$$

The meaning of the operator

$$\left(h \frac{\partial}{\partial x} + k \frac{\partial}{\partial y} \right)^{m} f(x, y)$$

is

$$\sum_{r=0}^{m} \binom{m}{r} h^{m-r} k^r \frac{\partial^m}{\partial x^{m-r} \partial y^r} f(x, y).$$

In the expansion of $f(a+h, b+k)$, the values (a, b) are inserted for (x, y) after performing the differentiations of orders up to $n-1$ and $(a+\theta h, b+\theta k)$ in the nth order terms. The hypothesis of continuity of the partial derivatives ensures that the order of differentiation in any

$$\frac{\partial^m}{\partial^{m-r} x \partial^r y} f(x, y) \quad (m \leqslant n)$$

is indifferent.

Proof of theorem 8.7. We reduce the number of variables to one (t) by writing $F(t) = f(a+ht, b+kt).$

By theorem 4.82,

$$F(1) = F(0) + F'(0) + \dots + \frac{1}{(n-1)!} F^{(n-1)}(0) + \frac{1}{n!} F^{(n)}(\theta),$$

where $0 < \theta < 1$.

The successive derivatives $F'(t)$, $F''(t)$, ... are calculated by means of theorem 8.5. With the prescribed notation,

$$F'(0) = \left(h\frac{\partial}{\partial a} + k\frac{\partial}{\partial b}\right) f(a, b),$$

$$F''(0) = \left(h^2\frac{\partial^2}{\partial a^2} + 2hk\frac{\partial^2}{\partial a\,\partial b} + k^2\frac{\partial^2}{\partial b^2}\right) f(a, b),$$

and so on. Finally, $F^n(\theta)$ has the value stated in the theorem. |

The possible extension of this result, on the lines of theorem 5.8, to provide an expansion of $f(x, y)$ in an infinite series of powers of x and y is less important than the expansion of $f(x)$ as $\Sigma a_n x^n$.

8.8. Maxima and minima

We extend to a function f of two independent variables the discussion of §7 of chapter 4.

Definition. f has a maximum at (a, b) if there is a neighbourhood of (a, b) in which $f(x, y) < f(a, b)$ except for $(x, y) = (a, b)$.

We define a *minimum* by substituting $>$ for $<$.

The following analogue of theorem 4.71 is immediate.

If f_x exists at (a, b), a necessary condition that f has a maximum or minimum at (a, b) is that $f_x(a, b) = 0$.

This follows by applying theorem 4.71 to $f(x, b)$.

A similar statement holds for $f_y(a, b)$.

The investigation of *sufficient* conditions for a maximum or minimum has a feature which was not present with only one independent variable. This is embodied in an algebraic lemma.

Lemma. If $\phi(x, y) = ax^2 + 2hxy + by^2$, where all the numbers are real, and $D = ab - h^2$, we have

(i) *if $D > 0$ and $a > 0$, then $\phi(x, y) > 0$ for all (x, y) except $(0, 0)$;*

if $D > 0$ and $a < 0$, then $\phi(x, y) < 0$ for all (x, y) except $(0, 0)$;

(ii) *if $D < 0$, there are values (x_1, y_1), (x_2, y_2) arbitrarily near to $(0, 0)$ for which*

$$\phi(x_1, y_1) > 0, \quad \phi(x_2, y_2) < 0.$$

Proof. (i) Note that $D > 0$ implies that $a \neq 0$. Then

$$a\phi = (ax+hy)^2+(ab-h^2)\,y^2$$

$$> 0 \quad \text{unless} \quad x = y = 0.$$

So, if $\qquad\qquad a > 0, \quad \text{then} \quad \phi > 0,$

and if $\qquad\qquad a < 0, \quad \text{then} \quad \phi < 0.$

(ii) Suppose that $a \neq 0$, say $a > 0$. Then, if $y_1 = 0$ and x_1 has any value (except 0), the expression for $a\phi$ in (i) gives

$$\phi(x_1, y_1) > 0.$$

If (x_2, y_2), not being $(0, 0)$, satisfy $ax_2+hy_2 = 0$, then

$$\phi(x_2, y_2) < 0.$$

If $a = 0$ and $b \neq 0$, a similar argument holds.

If $a = b = 0$, then $h \neq 0$ and $\phi(x, y) = 2hxy$, which takes opposite signs when $x = y$ and $x = -y$.

Theorem 8.8. *Suppose that f has derivatives of the first and second orders f_x, f_y, f_{xx}, f_{xy}, f_{yy}, which are continuous at (a, b), and write p, q, r, s, t for their values at (a, b).*

Then f has a maximum or minimum at (a, b) if

$$\text{(i)} \quad p = q = 0,$$

$$\text{(ii)} \quad rt-s^2 > 0.$$

If $r < 0$, (a, b) is a maximum, if $r > 0$ a minimum. (From **(ii)** *$r \neq 0$.)*

If $rt-s^2 < 0$, (a, b) is neither a maximum nor a minimum of f.

If $rt-s^2 = 0$, we prove nothing.

Proof. By theorem 8.7, if h, k are small enough,

$$f(a+h, b+k) = f(a, b)+ph+qk+\tfrac{1}{2}(r_1h^2+2s_1hk+t_1k^2),$$

r_1, s_1, t_1 being the values of f_{xx}, f_{xy}, f_{yy} for $x = a+\theta h, y = b+\theta k$, where θ, depending on a, b, h, k, satisfies $0 < \theta < 1$.

If condition (i) is satisfied,

$$f(a+h, b+k)-f(a, b) = \tfrac{1}{2}(r_1h^2+2s_1hk+t_1k^2).$$

It is now clear what we have to do to complete the proof.

The coefficients r_1, s_1, t_1 of the quadratic form in h, k tend to r, s, t as $(h, k) \to (0, 0)$, and we need to show that, for all sufficiently small (h, k), the two forms

$$r_1 h^2 + 2s_1 hk + t_1 k^2,$$
$$rh^2 + 2shk + tk^2,$$

both have a fixed sign or are both capable of taking either sign.

Write
$$D = rt - s^2, \quad D_1 = r_1 t_1 - s_1^2$$

(so that D_1 depends on h, k).

Suppose that $D > 0$. Then $r \neq 0$. Suppose that $r > 0$. Since $|r - r_1|$, $|s - s_1|$, $|t - t_1|$ are arbitrarily small if (h, k) is near enough to $(0, 0)$, we can choose δ such that, for all h, k in $h^2 + k^2 < \delta^2$,

$$D_1 > 0, \quad r_1 > 0.$$

By the lemma, for all h, k in $0 < h^2 + k^2 < \delta^2$,

$$r_1 h^2 + 2s_1 hk + t_1 k^2 > 0,$$

and so (a, b) is a minimum of f.

Similarly, if $D > 0$ and $r < 0$, then (a, b) is a maximum.

Take now the case $D < 0$.

By the lemma, there is (h_1, k_1) for which

$$rh^2 + 2shk + tk^2$$

is less than 0 and (h_2, k_2) for which it is greater than 0.

Write
$$F(t) = f(a + ht, b + kt) - f(a, b).$$

Then
$$F(0) = F'(0) = 0,$$
$$F''(0) = rh^2 + 2shk + tk^2.$$

For $h = h_1$, $k = k_1$, the function $F(t)$ has a maximum at $t = 0$ (by theorem 4.72). Hence, in every neighbourhood of (a, b), there is a point (x, y) for which $f(x, y) < f(a, b)$.

Similarly, from $h = h_2$, $k = k_2$, we arrive at a point for which $f(x, y) > f(a, b)$. |

8.9. Implicit functions

The problem can be introduced by a particular example. Suppose that
$$x^2 + y^2 = 1.$$

In what sense (if any) does this equation define y as a function of x? If a value of x is assigned in $-1 < x < 1$, there are two values of y which satisfy the equation. We stipulated on p. 47 that our functions should be single-valued. Suppose that we start from a point, say $(0, +1)$ and let x take values varying continuously. Then plainly we can find corresponding values of y such that (x, y) satisfies the equation and y varies continuously. In graphical language we must keep the point (x, y) on the upper part of the circle and not allow it to jump to the lower part.

Moreover, with the understanding that

$$y = +\sqrt{(1 - x^2)},$$

y has a derivative given by

$$y' = -\frac{x}{\sqrt{(1 - x^2)}}.$$

It is often inconvenient or impossible to solve an equation $F(x, y) = 0$ explicitly for y. The following theorem gives conditions under which we can nevertheless assert that the equation defines a function $y = f(x)$ and, further, that the function can be differentiated.

Theorem 8.9. *Suppose that, for (x, y) in a square with centre (a, b), $F(x, y)$ satisfies the conditions*
   (i) *$F(x, y)$ is differentiable,*
   (ii) *$F(a, b) = 0$,*
   (iii) *$F_y(x, y) > 0$.*
Then there is a function $y = f(x)$ defined in an interval with centre a such that
   (iv) *$y = f(x)$ satisfies $F(x, y) = 0$ identically,*
   (v) *dy/dx exists and is given by*

$$F_x + F_y \frac{dy}{dx} = 0.$$

Proof. In fig. 5, P_0 is the point (a, b) and conditions (i)–(iii) hold in the square shown. From condition (iii), $F(x, y) > 0$ at Q_0 and $F(x, y) < 0$ at R_0.

From (i), F is continuous, and so we can find vertical lines

$Q_1 R_1$ and $Q_2 R_2$ such that $F > 0$ for all points on $Q_1 Q_2$ and $F < 0$ for all points on $R_1 R_2$.

Let now QR be any vertical line between $Q_1 R_1$ and $Q_2 R_2$. By (iii), F increases (strictly) from a negative value at R to a positive value at Q. Since F is continuous, there is a unique point P at which $F = 0$. The ordinate of P defines $y = f(x)$, and (iv) is established.

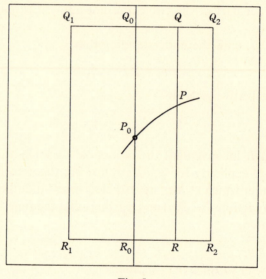

Fig. 5

We now prove (v). If (x, y) and $(x+h, y+k)$, where $h \neq 0$, are two points at which $F = 0$, we have, F being differentiable,

$$0 = F(x+h, y+k) - F(x, y)$$
$$= h F_x(x, y) + k F_y(x, y) + \epsilon(|h| + |k|),$$

where ϵ can be made arbitrarily small by taking (h, k) near enough to $(0, 0)$.

Divide through by h and (as we may by (iii) of the hypothesis) by $F_y(x, y)$. Then let $h \to 0$. It follows that $\lim (k/h)$ exists, and we have

$$\frac{F_x}{F_y} + \frac{dy}{dx} = 0. \quad \blacksquare$$

Exercises 8 (d)

Notes on these exercises are given on p. 183.

1. Prove that, if A, B, C are the angles of a triangle

$$\cos A + 2^{\frac{1}{2}}(\cos B + \cos C) \leqslant 2.$$

2. Investigate maxima and minima of

$$\text{(i)} \quad (x-y)^2 \, (a^2 - x^2 - y^2),$$

$$\text{(ii)} \quad (x-y^2) \, (x-2y^2).$$

3. If $f(x, y) \equiv x^3 + 2x^2y - xy^2 - 8y^3$, show that there is only one point (x_0, y_0) where necessary conditions for a maximum or minimum are satisfied. By considering the values of f on the line $y = y_0$, prove that it has no maximum or minimum.

4. If
$$u = x^2 + y^2, \quad v = x^3 + y^3,$$

prove that, if x is considered as a function of u, v,

$$\frac{\partial x}{\partial u} = -\frac{y}{2x(x-y)}, \quad \frac{\partial x}{\partial v} = \frac{1}{3x(x-y)}.$$

Prove that
$$\frac{\partial(x, y)}{\partial(u, v)} = -\frac{1}{6xy(x-y)}.$$

5. If $f(x, y, z) = 0$, where f is a differentiable function of x, y, z, prove that

$$\left(\frac{\partial z}{\partial y}\right)_x \left(\frac{\partial x}{\partial z}\right)_y \left(\frac{\partial y}{\partial x}\right)_z = -1,$$

where $(\partial z / \partial y)_x$ denotes the derivative of z with respect to y when x is constant.

6. The equations
$$f(x, y, z, w) = 0,$$
$$g(x, y, z, w) = 0,$$

where f and g are differentiable, can be solved to give z and w as functions of x, y. Prove that

$$\frac{\partial z}{\partial x} = -\frac{\partial(f, g)}{\partial(x, w)} \bigg/ \frac{\partial(f, g)}{\partial(z, w)}, \quad \frac{\partial z}{\partial y} = -\frac{\partial(f, g)}{\partial(y, w)} \bigg/ \frac{\partial(f, g)}{\partial(z, w)}.$$

Calculate $\partial z / \partial x$, $\partial z / \partial y$ as functions of x, y, z, w if

$$xy + zw = 0,$$
$$x^2 + y^2 - z^2 - w^2 = 1.$$

NOTES ON THE EXERCISES

Answers or hints for solution are given to those exercises in which they appear to be most useful. If an exercise embodies an important result or an instructive method, the solution is set out in more detail.

1 (a)

1. Write $l(n)$, $r(n)$ for the expressions on left and right sides. Assume, for some n, $l(n) = r(n)$. Then

$$l(n+1) = l(n)+(n+1)^2 = r(n)+(n+1)^2$$
$$= \tfrac{1}{6}(n+1)(n+2)(2n+3) = r(n+1).$$

But $l(1) = 1 = r(1)$. Hence $l(n) = r(n)$ for all n.

2. Right-hand side can be found from **1**, since

$$\text{left-hand side} = \sum_1^{2n} r^2 - 4 \sum_1^{n} r^2.$$

3. Put $n = 1, 2, 3$ to find A, B, C.

5. $l(n+1) > r(n+1)$ follows from $l(n) > r(n)$ if $2 > \{(n+1)/n\}^3$, i.e. if $n > 3$. $l(n) > r(n)$ is false if $n = 9$, true if $n = 10$.

7. $m = pd-qc$, $n = qa-pb$ will do.

1 (b)

3. Suppose that a/b is a root, where a and b are integers without a common factor, and $b > 0$. Write a/b for x and multiply by b^{n-1}. This gives $b = 1$.

4. ± 2, -3, 4; -1, $\tfrac{3}{2}$, $\tfrac{3}{2}$.

1 (c)

1. $99/70$, $239/169$.

3. $(a-c+\sqrt{b})^2 = d$ gives $a = c$ or $\sqrt{b}$ rational.

6. $ad = bc$.

1 (d)

3. If A and B are sets of positive numbers, then $\sup D = \sup A \sup B$ and $\inf D = \inf A \inf B$. If A or B may contain negative numbers, there is no such simple result.

5. If the a's are not all equal, there are two a_r, a_s for which $a_r < A < a_s$. Replace a_r, a_s by b_r, b_s where $b_r = A$, $b_s = a_r+a_s-A$. Then

$$b_r b_s - a_r a_s = (A-a_r)(a_s-A) > 0.$$

So the replacement has kept the A.M. the same and increased the G.M. After at most $n-1$ repetitions of the argument all the numbers are replaced by A. The G.M., now equal to A, having been increased by replacements was at first less than A.

6. Use the identity
$$(\Sigma a_r b_r)^2 = \Sigma a_r^2 \Sigma b_r^2 - \Sigma (a_r b_s - a_s b_r)^2,$$
where r, s take the values, $1, 2, \ldots, n$.

7. Prove that, for the positive integer n,
$$\frac{a^{n+1}-1}{n+1} > \frac{a^n-1}{n}.$$

Dividing by the positive factor $a-1$ and then multiplying up, we have proved this inequality if
$$na^n > a^{n-1}+a^{n-2}+\ldots+1,$$
which is true.

It follows that, if $m > n$,
$$\frac{a^m-1}{m} > \frac{a^n-1}{n}.$$

To extend to rational indices, let $r = m/p$ and $s = n/p$ (with m, n, p positive integers). Put $a^{1/p} = b$.

8. Similar to **7.**

1 (e)

1. $a-c = i(d-b)$ gives $(a-c)^2 = -(d-b)^2$. From the order properties of real numbers, left-hand side $\geqslant 0$ and right-hand side $\leqslant 0$.

3. $(a+bi)(c+di) = 0$ gives $ac-bd = ad+bc = 0$. Hence
$$(a^2+b^2)(c^2+d^2) = (ac-bd)^2+(ad+bc)^2 = 0.$$
Therefore $\qquad\qquad a^2+b^2 = 0 \quad\text{or}\quad c^2+d^2 = 0.$

1 (f)

2. Circle, line, circle, hyperbola foci ± 1, contradicts theorem 1.10 (2).

3. (iii) Combining the conjugate $2\bar{z}^3+z^3 = 3$ with the original equation we have $z^3 = \bar{z}^3$ and then $z^3 = 1$.

4. Put $z+p = Z$.

5. If $|z| \leqslant \frac{1}{3}$, then $|a_1 z+\ldots+a^n z_n| \leqslant 2\{\frac{1}{3}+\ldots+(\frac{1}{3})^n\} < 1$.

6. Circle.

8. If there is a root with $|z| = 1$, then $\bar{z} = 1/z$, and so $c\bar{z}^2+b\bar{z}+a = 0$. Conjugate gives $\bar{c}z^2+\bar{b}z+\bar{a} = 0$. Combine with original equation to give
$$z(ab-b\bar{c})+a\bar{a}-c\bar{c} = 0.$$
For converse, write $w = (\bar{b}c-\bar{a}b)/(a\bar{a}-c\bar{c})$, so that $|w| = 1$. $aw+b$ reduces to $-c\bar{w}$. But $\bar{w} = 1/w$.

9. Roots of quadratic are $v+v^2+v^4$ and its conjugate $v^6+v^5+v^3$. Form sum and product.

12. Prove first for a $P(z)$.

13. Use **12.**

2 (a)

1. $\frac{1}{2}+\frac{1}{4}(-1)^n$.

5. $s_n = 1/\{\sqrt{(n+1)}+\sqrt{n}\} \to 0$.

8. If $s \neq s'$, take $\epsilon = \frac{1}{2}|s-s'|$. $\exists N$. $|s_n-s| < \epsilon$ and $|s_n-s'| < \epsilon$ for all $n > N$. Contradiction.

2 (b)

8–10. All false. Counterexample for **9**, $s_n = n^2+(-1)^n n$.

2 (c)

2. Limit 0, a_0/b_0, $+\infty$, $-\infty$ according as $p < q, p = q, p > q$ and a_0/b_0 positive, $p > q$ and a_0/b_0 negative.

2 (d)

6. By the binomial theorem

$$s_n = \left(1+\frac{1}{n}\right)^n = 1+n\,\frac{1}{n}+\frac{n(n-1)}{1.2}\frac{1}{n^2}+\ldots+\frac{1}{n^n}.$$

We shall prove s_n increasing and bounded. The $(r+1)$th term of the binomial expansion, namely

$$\frac{1}{1.2\ldots r}\left(1-\frac{1}{n}\right)\ldots\left(1-\frac{r-1}{n}\right)$$

increases as n increases. Moreover, the number of terms increases with n. Therefore s_n increases.

From the above expansion

$$s_n < 1+1+\frac{1}{2!}+\frac{1}{3!}+\ldots+\frac{1}{n!}$$

$$< 1+1+\frac{1}{2}+\frac{1}{2^2}+\ldots+\frac{1}{2^n} < 3.$$

By theorem 2.6, s_n tends to a limit e, where $2 < e \leqslant 3$.

7. $^{n+1}\sqrt{(n+1)} < \sqrt[n]{n}$ if $(n+1)^n < n^{n+1}$, i.e. if $(1+n^{-1})^n < n$ which is true if $n \geqslant 3$ (proved in last exercise).

Write $\sqrt[n]{n} = 1+x$. Then $n = (1+x)^n > 1+\frac{1}{2}n(n-1)x^2$, and so $x^2 < 2/n$. Hence $x \to 0$ as $n \to \infty$.

11. $u_n = 0$ if $x = 0$ or m a positive integer. Otherwise $|u_{n+1}/u_n| \to |x|$.

12. (i) 3. (ii) 1, 0, -1 according as $a >, =, < b$.

(iii) $n\,\dfrac{n}{n^2+1} > s_n > n\,\dfrac{n}{n^2+n}$.

(iv) $n! > a^n$ for $n > N$; $\sqrt[n]{(n!)} \to \infty$.

(v) Method of **2 (a), 5.** Limit is $-\frac{1}{2}(a+b)$.

(vi) Use **2 (d), 1.**

2 (e)

By methods of §8, limits are

1. 4.   **2.** 1.   **3.** The smaller root.

4. 2.   **5.** 3, but s_4 is undefined if $s_1 = \frac{5}{7}$.

6. 2.

7.
$$l = \tfrac{1}{2}(1+\sqrt{21}), \quad u_{n+1}^2 - l^2 = u_n - l,$$

$$u_{n+1} - l = (u_n - l)/(u_{n+1} + l) < (u_n - l)/2l, \quad 4l^2 > 30.$$

8. 1, 2, 3.

9.
$$a_n > a_{n+1} > b_{n+1} > b_n \quad \text{and} \quad a_{n+1} - b_{n+1} < \tfrac{1}{2}(a_n - b_n).$$

Also
$$a_{n+1}b_{n+1} = a_n b_n.$$

2 (f)

1. $1/n^2 < 1/(n-1)n$ gives

$$\sum_{N+1}^{\infty} \frac{1}{n^2} < \frac{1}{N}.$$

Hence $N = 10^4$.

For $\Sigma\, (0\cdot 99)^n$, we have from $(0\cdot 99)^N < 10^{-4}(1 - 0\cdot 99)$, $N > 1376$.

The second series converges more rapidly than the first.

2. First is G.P. convergent for $r > 0$ to sum $1 + r$; also for $r = 0$ to sum 0. For third, if

$$s_n = \sum_{m=0}^{n} m r^{m-1},$$

$$s_n(1-r) = \{(1 - r^n)/(1-r)\} - nr^n.$$

3. Less than 3 by about $2\cdot 7 \times 10^{-11}$.

5. $\dfrac{1}{n(n+1)(n+2)} = \dfrac{1}{2}\left\{ \dfrac{1}{n(n+1)} - \dfrac{1}{(n+1)(n+2)} \right\}.$

2 (g)

1. *General.* If A, B are statements, the notation $A \Rightarrow B$ means 'If A, then B'. We say also that (i) A implies B, or (ii) A is a sufficient condition for B, or (iii) B is a necessary condition for A.

The double-headed arrow $A \Leftrightarrow B$ means that A and B are logically equivalent to each other, in other words: (i) If and only if A, then B; (ii) A is a necessary and sufficient condition for B.

In the text we shall nearly always write words instead of the symbols $\Rightarrow$, $\Leftrightarrow$.

In the particular examples in **1**, the conditions are (i) N and S, (ii) N not S, (iii) N only if $p \neq 0$, not S.

2. (i) N, not S, (ii) S, not N, (iii) N and S.

3. (i) $u_n \to \tfrac{1}{2}$; (ii) G.P.; (iii) s_n (n odd, n even); (iv) $1/n! < 1/2^{n-1}$.

5. Cf. 2 (f), 5.

6. For $s_n < s_{n+1}$, see **2** (*d*), **6**. To prove

$$\left(1-\frac{1}{n}\right)^n < \left(1-\frac{1}{n+1}\right)^{n+1}$$

take $a_1 = a_2 = \ldots = a_n = 1-(1/n)$, $a_{n+1} = 1$ and use

G.M. of $(a_1, \ldots, a_{n+1})$ < A.M.

$$t_n - s_n = \{1-(1-n^{-2})^n\}/(1-n^{-1})^n.$$

To prove numerator $\to 0$, use $(1-k)^n > 1-nk$.

7. $s_n = (-1)^n$, $t_n = (-1)^n$ or $(-1)^{n+1}$.

8. $6(n^3-1)/9(n-1)n(n+1)$.

10. If $r_n \to l$, l satisfies $l+(1/l) = 2A$. If roots of this are real, $A \geqslant 1$. If c is the larger root, $r_n - c = (r_{n-1}-c)/r_{n-1}c$, and induction.

11. A sequence oscillating more slowly as n gets larger, like

$$s_n = 2 + \sin(\pi\sqrt{n}).$$

If $s_{n+1}/s_n \to -\frac{1}{2}$, then $s_n \to 0$.

12. Last part, $nu_n^2 < (n+\frac{1}{2})u_n^2 < (1+\frac{1}{2})u_1^2$.

13. (i) True. (ii) False; we can say only $s \geqslant t$. Illustration of $s = t$ given by $-t_n = s_n = 1/n$. (iii) Falsity shown by $s_n = n^2$ (*n* even), $s_n = (n-1)^2$ (*n* odd); then $s_n \to \infty$.

14. Write $t_n = (s_1+s_2+\ldots+s_n)/n$.

Given ϵ, $\qquad s-\epsilon < s_n < s+\epsilon$ for $n > m$.

Sum from $m+1$ to n

$$(n-m)(s-\epsilon) < nt_n - mt_m < (n-m)(s+\epsilon).$$

Divide by n and rearrange

$$\left(1-\frac{m}{n}\right)(s-\epsilon)+\frac{m}{n}\,t_m < t_n < \left(1-\frac{m}{n}\right)(s+\epsilon)+\frac{m}{n}\,t_m.$$

Keeping m fixed, we can choose n_0 such that the first expression is greater than $s-2\epsilon$ and the last less than $s+2\epsilon$.

So $\qquad s-2\epsilon < t_n < s+2\epsilon$ for $n > n_0$.

15. Sum of n terms is

$$\frac{1-x^n}{(1-x)^2} - \frac{nx^n}{1-x}.$$

3 (*b*)

4. It is assumed (anticipating chapter 6) that the sine is continuous. If $x \neq 0$, $\sin(1/x)$ is continuous by theorem 3.5, and $f(x)$ is the product of two continuous functions. Continuity at $x = 0$ follows from the definition; $|f(x)| < \epsilon$ if $|x| < \epsilon$.

5. If $k = 1$, there is only one value of x.

If $k \neq 1$, $f(x) = k$ gives
$$(1-k)x^2 - (6-9k)x + 5 - 18k = 0.$$

This gives real x if $(6-9k)^2 \geqslant 4(1-k)(5-18k)$, i.e. if $9k^2 - 16k + 16 \geqslant 0$ which is true for all k.

8. (i) $x-1$ is a common factor.

9. (i) Multiply numerator and denominator by $\sqrt{(1+x)} + \sqrt{(1-x)}$.

(iii) Case $p < q$. If $q-p$ is even, limit $+\infty$ or $-\infty$ according as $a_0/b_0 > $ or < 0. If $q-p$ is odd, consider $x \to 0+$ and $x \to 0-$ separately.

10. (i) We want
$$\left(p + \frac{q}{x} + \frac{r}{x^2}\right)\left(1 + \frac{1}{x} + \frac{1}{x^2}\right)$$

to be as close as possible to $1/x$ for large x. Equate coefficients of $1/x^k$ for $k = 0, 1, 2$. $p = 0, q = 1, r = -1$.

(ii) As in (i), squaring to get rid of $\sqrt{}$. Or, anticipating §5·8, expand $\{1 + (4/x^2)\}^{-\frac{1}{2}}$.

3 (c)

1. (i) Continuous if x is irrational or 0.

(ii) Discontinuous for $x = n\pi$ $(n \neq 0)$.

(iii) Discontinuous for $1/(x-a) = n\pi$.

2. $f(x) = 1 - x$ for $0 < x < 1, f(0) = 0, f(1) = 1$ or $g(x) = x$ (x rational), $g(x) = 1 - x$ (x irrational).

4. $g(x)$ of **2** is continuous for $x = \frac{1}{2}$ only.

5. 10^{-3}.   **6.** (i) $f(0) = 0$,   (iii) $f(0) = a$.

7.
$$f(x+\delta) - f(x) \leqslant \tfrac{1}{2}\{f(x+2\delta) - f(x)\}\ldots$$
$$\leqslant \frac{1}{2^n}\{f(x+2^n\delta) - f(x)\},$$

where $a < x < x + 2^n\delta < b$. Let $\delta \to 0$ and $n \to \infty$.

8. $|x_n - x_{n-1}| \leqslant \alpha^{n-1}|x_1 - x_0|$.

$\sum\limits_{1}^{\infty}(x_n - x_{n-1})$ is (absolutely) convergent, i.e. $x_n \to \xi$.

$$|\xi - f(0)| = |f(\xi) - f(0)| \leqslant \alpha|\xi|$$

gives $-\alpha\xi \leqslant \xi - f(0) \leqslant \alpha\xi$, etc.

4 (a)

1. (i) $y = 4(x-2), y = -4(x+2)$. (ii) $y = -4$.

2. $y - 2 = -\frac{1}{2}x, y - 2 = \frac{1}{2}(x+4)$.

3. $y = 3x$.   **5.** $y = |x-1| + |x+1|$.

6. (i) (a) and (b) $x = $ integer.   (ii) (b) $x = 1$.

7. (i) Touch at $(0, 0)$, arc tan 18 at $(\frac{3}{2}, \frac{9}{8})$.

(ii) $\frac{1}{2}\pi$ at both points.

4 (b)

9. Write $p(x) = (x-a)^m (x-b)^n q(x)$, where a and b are consecutive roots of $p(x) = 0$. Prove that, if $p'(x) = (x-a)^{m-1} (x-b)^{n-1} r(x)$, then $r(a)$ and $r(b)$ have opposite signs.

10. If $x \neq 0, f'(x) = 2x \sin(1/x) - \cos(1/x)$, by the rules of §2.
  If $x = 0, \{f(h)-f(0)\}/h = h \sin(1/h) \to 0$, giving $f'(0) = 0$.

Examples like this are repeatedly used to settle some of the less easy questions which occur in differentiation. This one gives the answer *Yes* to the question (in geometrical language)—can a curve have a tangent at every point and yet the direction of the tangent not vary continuously?

12. Differentiate four times, using **11** and rejecting vanishing determinants.

4 (c)

1, 2, 5. Put into partial fractions.

4. $\sin 3x \sin 5x = \frac{1}{2}(\cos 2x - \cos 8x)$, and use **3**.

6. The only way of doing this systematically is to use complex partial fractions

$$\frac{a}{a^2+x^2} = \frac{1}{2i}\left(\frac{1}{x-ia} - \frac{1}{x+ia}\right).$$

The nth derivative is

$$\frac{(-1)^n n!}{2i}\left\{\frac{1}{(x-ia)^{n+1}} - \frac{1}{(x+ia)^{n+1}}\right\}.$$

If $r = \sqrt{(x^2+a^2)}$ and $\cos\theta = x/r$, $\sin\theta = a/r$, then de Moivre's theorem (**1 (e), 5**) gives

$$(x+ia)^{n+1} = r^{n+1}\{\cos(n+1)\,\theta + i\sin(n+1)\theta\},$$

etc. The result finally takes the real form

$$(-1)^n n!\, r^{-(n+1)} \cos(n+1)\theta,$$

where r, θ have the assigned values.

4 (d)

2. Method of **4 (b), 9**.

5. At most one real root of $p(x) = 0$ lies between two consecutive roots of $p'(x) = 0$.
  In the example, if n is odd, $p'(x) = -(1+x^n)/(1+x) < 0$ for all x. $p(x)$ decreases as x increases from large $-$ to large $+$ values. $p(x) = 0$ for one x.
  If n is even, $p'(x) < 0$ if $x < 1$, $p'(x) > 0$ if $x > 1$ and $p(1) > 0$.

6. Use first part of **5**.

4 (e)

2. Limit $\frac{1}{2}$.

3. We may take $l = 0$ (and apply the special case to $f(x) - lx$).
  Given ϵ, choose X to make $|f'(x)| < \epsilon$ for $x > X$.

$$f(x) - f(X) = (x-X)f'(c).$$

Divide by x. Choose X_1 to make $|f(X)| < \epsilon X_1$. Then

$$-2\epsilon < f(x)/x < 2\epsilon \quad \text{if} \quad x > X_1.$$

4. The discontinuous $f'(x)$ in **4 (b), 10** shows that there is a false step in the argument.

4 (f)

1. Put $x = \frac{1}{2}+h$, or use §4.9.   (i) $(m/n)\,(\frac{1}{2})^{m-n}$; (ii) $\frac{1}{2}\pi^2$.

2. $a+b$.

3.
$$y-\frac{3at^2}{1+t^3} = \frac{t(2-t^3)}{1-2t^3}\left(x-\frac{3at}{1+t^3}\right).$$

If this meets the curve in point with parameter u, the product of the roots of the cubic in u is -1, and so $u = t, t$ or $-1/t^2$. As $t \to -1$, tangent approaches the asymptote $x+y+a = 0$.

5. (i) Maxima where $x^2-2x+3 = 0$, minimum at $x = 1$. (ii) None.

6. Greatest $3\sqrt{3}$ for $x = \frac{1}{3}\pi$; least 0.

8. dy/dx must not change sign, $b^2 \leqslant 3ac$.

9. $y = k$ touches the curve $y = p(x)/q(x)$.

10. Hemisphere.

11. Divide line of centres in the ratio $a^{3/2}$ to $b^{3/2}$.

12. Volume is greatest when P, Q, R are mid-points of BC, CA, AB.

13. $0 < k < 1$.

16, 17. Induction.   **18.** $n = 0$ and Leibniz.

19. $\exists x_1.\ a < x_1 < c, f'(x_1) > 0$.

20–22. Use §4.9.   **21.** (i) $\frac{1}{3}$; (ii) $\frac{1}{2}$.   **22.** $\frac{1}{2}n(n+1)$.

24. Lef $f'(a) = \alpha, f'(b) = \beta$ and suppose $\alpha < \gamma < \beta$. Write

$$g(x) = f(x)-\gamma(x-a).$$

g has a minimum for some c between a and b, and $g'(c) = 0$.

5 (a)

1. C, C, C for $x \leqslant 1$, C for $x < 4$, C for $x < \frac{3}{2}$.

5. $b^n/a^n \to \infty$, $a^{n+1}/b^n \to 0$.

6. If $u_n = n^{-k}$, then $\sqrt[n]{u_n} \to 1$ and $u_{n+1}/u_n \to 1$.

8. (i) T; (ii) F; (iii) F (see 3).

5 (b)

2. $k > 1, k > 0$.

3. (i) None; (ii) -1.

4. (i) $|u_n| \to \frac{1}{4}$, therefore D; (ii) abs. C; (iii) C by theorem 5.22.

5. $a = b$, C. $a \neq b$, D.

6. Sums of 5 and 6 terms are $\frac{37}{60}$ and $\frac{47}{60}$.

7. (i) The hypothesis in theorem 5.22 that a_n decreases is omitted. To prove the statement false, consider

$$a_n = 1/n^{\frac{1}{2}} + (-1)^n/n.$$

(ii) $|u_n| \leqslant A |v_n|$. Hence $\Sigma |u_n|$ converges.

5 (c)

1. See 2 (f), 2.

2. $|z| < |1 - z|$.

3. (i) Theorem 5.22 to re and im parts. (ii) $|u_n| = 2^{\frac{1}{2}n} n^{-2} \to \infty$. (iii) u_n is $1/n$ or i/n according as n is even or odd.

5 (d)

1. (i) 1, (ii) all z, (iii) e, (iv) 0, (v) 2, (vi) 1, (vii) 1, (viii) $\frac{1}{4}$.

2. $a < 1$, C all z.

5. $R \geqslant 1$, $R = 1$.

6. Use theorem 1.10 (sum).

7. If $r > s$, radius of convergence is s. If $r = s$, any number $\geqslant r$ (give illustrations).

8. $(1 - z) \sum_1^m \dfrac{z^n}{n} = z - \sum_1^{m-1} \dfrac{z^{n+1}}{n(n+1)} - \dfrac{z^{m+1}}{m}$.

For $|z| = 1$, this tends to a finite limit.

9. $R = \frac{3}{2}$. If $|z| = \frac{3}{2}$, modulus of nth term > 1, and so series diverges on the circle.

5 (e)

1. (i) C; (ii) C if $a < \max(b, c)$; (iii) C if $k > \frac{1}{2}$; (iv) C if $k > 0$; (v) D; (vi) C.

2. $\dfrac{1}{a-1} - \dfrac{1}{a+1} = \dfrac{2}{a^2-1}$.

3. $(n-m)u_n < \sum_{m+1}^n u_r < \sum_{m+1}^\infty u_r \to 0$ as $m \to \infty$. Put $m = \frac{1}{2}n$ (if n even) or $\frac{1}{2}(n+1)$ (if n odd).

4. Of terms with m-digit denominators, the number remaining is $8 \times 9^{m-1}$. Sum is less than

$$8 \left(\frac{1}{1} + \frac{9}{10} + \frac{9^2}{10^2} + \ldots + \frac{9^{m-1}}{10^{m-1}} + \ldots \right) = 80.$$

5. Theorem 5.7.

$$6\,(a)$$

1. If $e = l/m$, then

$$m!\left(\frac{l}{m}-\overset{m}{\underset{0}{\Sigma}}\frac{1}{n!}\right) = m!\overset{\infty}{\underset{m+1}{\Sigma}}\frac{1}{n!} < \frac{1}{m},$$

and so an integer is less than $1/m$. Contradiction.

2. Included in **3**.

3. The argument of **2 (d), 6,** shows that, if $x > 0$,

$$\left(1+\frac{x}{n}\right)^{n} < \exp x < \left(1-\frac{x}{n}\right)^{-n}.$$

(The inequalities for $x < 0$ will then follow by putting $x = -y$.)
  Now we prove that

$$\left(1-\frac{x}{n}\right)^{-n}-\left(1+\frac{x}{n}\right)^{n} \to 0.$$

$$\text{L.H.S.} = \left(1+\frac{x}{n}\right)^{n}\left\{\left(1-\frac{x^2}{n^2}\right)^{-n}-1\right\}$$

$$< \exp x \left\{\left(1-\frac{x^2}{n}\right)^{-1}-1\right\} = \frac{x^2 \exp x}{n-x^2}$$

(by use of easy inequality

$$\{1-(b/n)\}^{-n} < (1-b)^{-1} \quad \text{when} \quad 0 < b < 1).$$

4. $f_n'(x) < 0$ for $x > 0$, and so f_n decreases from $1-k$ to $-k$.

$$f_{n+1}(x) = f_n(x)+\frac{x^{n+1}e^{-x}}{(n+1)!} \quad \text{gives} \quad f_{n+1}(x_n) > 0.$$

5. 9120.   **6.** Descending order as given.

7. $\exp(\sqrt{x})$.

$$6\,(b)$$

1. As $h \to 0$,

$$\frac{x^h-1}{h} = \frac{\exp(h\log x)-1}{h} \to \log x.$$

2. The derivatives satisfy

$$1-x < (1+x)^{-1} < 1-x+x^2.$$

3. $yf'(xy) = f'(x)$ and $xf'(xy) = f'(y)$.

4. If $x = (y-1)/(y+1)$, L.H.S. $= \log(1+x)-\log(1-x)$.

5. (i) $a, 0$; (ii) $\log(a/b)/\log(c/d)$. For $x \to \infty$, several cases, e.g. if max (a, b, c, d) is a, limit is ∞.

$$6\,(c)$$

1. Take four terms of series for $\cos x$ when $x = \frac{3}{2}$ and three terms when $x = \frac{8}{5}$.

2. From theorem 6.81, $\sin x > 0$ for $0 < x \leqslant \frac{1}{2}\pi$ and from 6.82 (1),

$\sin x > 0$ for $\frac{1}{2}\pi < x < \pi$. From 6.82 (2), $\sin x < 0$ for $\pi < x < 2\pi$. So $\sin (x+c) = \sin x$ cannot be true for all x if $c < 2\pi$.

4. $(\log \cos \alpha y)/y^2 \to -\frac{1}{2}\alpha^2$ (e.g. by §4.9). Put $y = 1/n$.

6 (d)

1. $\cosh (\alpha + n\beta) \sinh (n+1) \beta \operatorname{cosech} \beta$.

2. Method of §2.8.

3. Real and imaginary parts of geometric series $\Sigma r^n \exp (in\theta)$.

4. Analogous formulae valid for $|r| < e^{-\theta}$ if $\theta > 0$.

5. Induction.

6. (i) Use $\cos 3x$.   (ii) $x + \frac{1}{3}x^3 + \frac{2}{15}x^5$.
   (iii) $(1 - x^2)y'' - xy' = 2$. Leibniz gives

$$(1 - x^2)y^{(n+2)} - (2n+1) xy^{(n+1)} - n^2 y^{(n)} = 0.$$

Putting $x = 0$, we obtain Maclaurin coefficients.
   (iv) Method of (iii). General term is

$$(-1)^n m(m^2 - 1^2) (m^2 - 3^2) \ldots \{m^2 - (2n-1)^2\} x^{2n+1}/(2n+1)!$$

   (v) $\exp (1+2i)x = \Sigma(1+2i)^n x^n/n!$
Define θ by $\cos \theta = 1/\sqrt{5}$, $\sin \theta = 2\sqrt{5}$.
Real part is $\Sigma 5^{\frac{1}{2}n} \cos n\theta(x^n/n!)$.
   (vi) $1 - \frac{1}{2}x^2 + \frac{1}{2}x^3$.

7 (a)

2. Take ξ_r to be left-hand end-point of δ_r.

$$\sum_{r=1}^{n} f(\xi_r)\delta_r = \sum_{r=1}^{n} (aq^{r-1})^k (aq^r - aq^{r-1})$$

$$= a^{k+1}(q-1) \Sigma q^{(r-1)(k+1)}$$

$$= a^{k+1}(q-1) \{q^{n(k+1)} - 1\}/(q^{k+1} - 1)$$

$$= (b^{k+1} - a^{k+1}) (q-1)/(q^{k+1} - 1).$$

As $n \to \infty$, $q \to 1$ and $(q^{k+1} - 1)/(q-1) \to k+1$.

5. Given ϵ, there is, by definition of J, a dissection $\mathcal{D}_0$ for which

$$J \leqslant S(\mathcal{D}_0) < J + \epsilon.$$

Let $\mathcal{D}_0$ have p points of division inside (a, b). Let $\mathcal{D}$ be any dissection, with norm δ^*, and $\mathcal{D}_1$ the dissection formed by all points of division of $\mathcal{D}_0$ and $\mathcal{D}$.

Then (theorem 7.21) $S(\mathcal{D}_1) \leqslant S(\mathcal{D}_0)$. Also, since $\mathcal{D}_1$ is formed from $\mathcal{D}$ by p extra points of division,

$$S(\mathcal{D}) - S(\mathcal{D}_1) \leqslant p(M-m)\delta^*.$$

Hence $\qquad S(\mathcal{D}) \leqslant J + \epsilon + p(M-m)\delta^*.$

If $\delta^* < \epsilon/(M-m)p$, then $J \leqslant S(\mathcal{D}) < J + 2\epsilon.$

7 (b)

Most of the exercises in this and the next set depend directly on the methods of §§7.7 and 7.8.

3. In $(0, \frac{1}{2}\pi)$, $\sin^{2m-1} x \geqslant \sin^{2m} x \geqslant \sin^{2m+1} x$,

giving $I_{2m-1} \geqslant I_{2m} \geqslant I_{2m+1}$ (in fact, >).

Divide by I_{2m+1} and use $(2m+1) I_{2m+1} = 2m I_{2m-1}$.

7 (c)

6. $2 \arctan \{(1+r) \tan \frac{1}{2}\delta/(1-r)\} + \delta$ $(r<1)$,

$-2 \arctan \{(r+1) \tan \frac{1}{2}\delta/(r-1)\} + \delta$ $(r > 1)$.

Limits $\pi + \delta$ $(r < 1)$, $-\pi + \delta$ $(r > 1)$. $I = \delta$ when $r = 1$.

7 (d)

2. First sum lies between

$$\int_n^{2n} \frac{dx}{x} \quad \text{and} \quad \int_{n+1}^{2n} \frac{dx}{x}$$

and so $\to \log 2$. Second sum $\to 0$ by theorem 5.22.

3–6 and **8** follow from theorem 7.11.

7. The idea of theorem 7.11 used and applied to an *increasing f*.

9. (i) By §6.6, if $\delta > 0$, $(\log n)^q/n^\delta \to 0$. If $p < 1$, choose $\delta = \frac{1}{2}(1-p)$, say. $\Sigma(1/n^{p+\delta})$ diverges. Therefore so does

$$\Sigma \frac{n^\delta}{n^{p+\delta} (\log n)^q}.$$

Similarly, convergence if $p > 1$. If $p = 1$, we have **6**.
   (ii) $(\log \log n)^q < \log n$ if $n > n_0$. Divergent for all q.

7 (e)

6. (a) Since $y'' < 0$, the curve $y = \log x$ is concave to Ox.
   (i) First inequality expresses that the area under the curve $y = \log x$ between $x = r-\frac{1}{2}$, $x = r+\frac{1}{2}$ is less than area of trapezium formed by $y = 0$, $x = r-\frac{1}{2}$, $x = r+\frac{1}{2}$ and tangent at $x = r$.

7 (f)

1. (i) $\sqrt{2}$, from theorem 7.62.   (ii) 1.

2. $g(x) = (x-1) \int_0^x \xi f(\xi) \, d\xi + x \int_x^1 (\xi-1) f(\xi) \, d\xi.$

Using theorem 7.62

$$g'(x) = \int_0^x \xi f(\xi) \, d\xi + \int_x^1 (\xi-1) f(\xi) \, d\xi.$$

$$g(0) = g(1) = 1.$$

5. (i) $\displaystyle\int_0^1 \frac{dx}{1+x^2}$;

(ii) $\log f(n) = \dfrac{1}{n}\left\{\log\left(1+\dfrac{1}{n}\right)+\ldots+\log\left(1+\dfrac{n}{n}\right)\right\} \to \displaystyle\int_0^1 \log(1+x)\,dx$.

6. Method of **5**.

7. Integrate by parts. Induction. For last part,

$$I_n(\tfrac{1}{2}\pi) < \int_{-1}^1 (1-x^2)^n\,dx < 2 \quad \text{and} \quad b^{2n+1}/n! \to 0$$

as $n \to \infty$, giving an integer equal to a fraction.

8 (a)

1. No, no, yes.

8 (b)

1. (i) Yes. (ii) If $k > \tfrac{1}{2}$.

2. Yes, yes, no.

3. (Cf. **3** (c), **4**.) If $f(x) = 0$ for x irrational and x^2 for x rational, $f'(x) = 0$ at $x = 0$ and f is discontinuous for all other x. To construct an analogous function $g(x, y)$ of two variables, define $g(x, y) = f(r)$ if $x^2 + y^2 = r^2$ ('surface of revolution').

8 (c)

1. (i)
$$\phi_u u_x + \phi_v v_x = 0,$$
$$\phi_u u_y + \phi_v v_y = 0$$

have solutions other than $(0, 0)$ for ϕ_u, ϕ_v.

(ii) Multiplication of determinants.

2. $u = f(x) + f(y)$, $v = (x+y)/(1-xy)$.

3.
$$\Delta = \tfrac{1}{2}a^2 \sin B \sin C/\sin A.$$

$$\frac{\delta\Delta}{\Delta} = \frac{2\delta a}{a} + \cot B\,\delta B + \cot C\,\delta C - \cot A\,\delta A$$

and
$$\delta A + \delta B + \delta C = 0.$$

4. Method of **3**. $\delta\Delta$ is greatest when

$$\delta a > 0,\ \delta b > 0,\ \delta c > 0 \quad \text{(acute-angled),}$$

or
$$\delta a < 0,\ \delta b > 0,\ \delta c > 0 \quad \text{(obtuse at } A\text{).}$$

5. $y = f(u) + g(v)$, where f, g are arbitrary (differentiable) functions.

6 (also **7, 9**). Transformation as in §8.6.

$$u^2 = x + \sqrt{(x^2+y^2)}, \quad g = Au + B.$$

8. As in §8.6.

8 (*d*)

1. Put $-\cos (B+C)$ for $\cos A$. Necessary condition for turning value is $B = C$. Maximum of $-\cos 2B + 2^{3/2} \cos B$ given by $B = \frac{1}{4}\pi$.

2. (i) Maximum at $(0, \pm a/\sqrt{2})$.

(ii) $f_x = f_y = 0$ at $(0, 0)$. This is not a maximum or minimum because, near $(0, 0)$, $f < 0$ between the parabolas $y^2 = x$, $2y^2 = x$ and > 0 elsewhere.

3. $(0, 0)$.

5. Method of **6**.

6. δx, δy, δz, δw satisfy (approximately)

$$f_x\, \delta x + f_y\, \delta y + f_z \delta z + f_w\, \delta w = 0$$

and similarly for g.

Put $\delta y = 0$, solve for $\delta z / \delta x$ and let $\delta x \to 0$.

$$\frac{xz + yw}{z^2 - w^2}, \quad \frac{xw + yz}{z^2 - w^2}.$$

INDEX